1

ABOUT THE AUTHOR

Steven Burr started life as a university student in 1990. Having now graduated from five different ivory towers, Steven thinks that he knows a thing or two about what students both want and need to know. Only now is he beginning to realize that these may actually have been two different things all along.

Steven hopes that you will find his textbook both interesting and useful. He also hopes that you will recommend it to all your friends, and that enough people will buy copies to make him a millionaire. He can always hope...

Steven worked for a short time as a forensic scientist and learnt two things. Firstly that forensic work is surprisingly very boring. Secondly that given half a chance many people will do some truly awful things to other peoples' bodies. This is why Steven quickly returned to a nice safe job in a cosy ivory tower. Steven agrees that he doesn't actually do any real work anymore, he just has 'hobbies'.

1005246000

Some people say that Steven looks like Ned Flanders. Steven simply responds by saying that looks can be deceiving, and those who know him well say that he is dead right. His research focuses on nerve poisons. He also has an interesting collection of surgical implants (well his wife says they're interesting!), and a fascination with the darker side of oriental culture.

First Edition
Published by S.A. Burr, Nottingham, September 2006.
Distributed by Nottingham University Press,
see www.nup.com or telephone +44 (0)115 9831001.
ISBN 0-9554151-0-1 / 978-0-9554151-0-4

THE BODY IN BALANCE
A PHYSIO-PHILOSOPHICAL VIEW OF LIFE

by **STEVEN ASHLEY BURR,** BSc (Hons). MSc. PGDE (FAHE). MMedSci (ClinEd). PhD. AIBMS. CBiol. MIBiol. ILTM.

PREFACE

This is yet another undergraduate textbook on basic biomedical science, the principal focus of which is human physiology.

However the format of the knowledge is innovative. The aim of the book is to categorize and quantify the different qualities that are the essential requirements for life, but in a new way using a process rather than a systems based approach. I have also attempted to integrate important insights regarding measurement and include as much clinical relevance as possible. Diagrams are aimed to be intuitive and are included without explanatory legends in order to encourage deeper thought of their relevance to the main body of the text. The study of life is more intriguing than fiction and here the reader should find some interesting ideas not found elsewhere. This is not intended to be a comprehensive textbook, but one that adopts a lateral approach to the subject and fills-in important gaps. For all of you who are students in this field, I hope this work will stimulate thought, help understanding, be a useful reference and aid revision.

ACKNOWLEDGEMENTS

Of the very many people who have helped me along the way, five standout for special mention. First of all, my parents for the sacrifices they have made for my education and their seemingly never-ending support.
My friend, Mr. Lee Craig Hampton (BA. PGCE. MA.) for his expert handling of the art and design work for this book. My MSc and PhD project supervisor Professor David Edward Ray (BSc. PhD.) for his kind and honest tutelage. Finally and most of all I thank my wife Yee-Ling Leung (BSc. MBChB. MRCS.) for editing the manuscript which became this book, and for all her love which makes my life worthwhile.

ABBREVIATIONS

c.   *circa*, meaning approximately

cf.   *confer*, meaning compare

e.g.   *exempli gratia*, meaning for example

etc.   *et cetera*, meaning and so forth

i.e.   *id est*, meaning that is

NB   *nota bene*, meaning note well

rf.   meaning refer to

vs.   *versus*, meaning in contrast to

CONTENTS

5

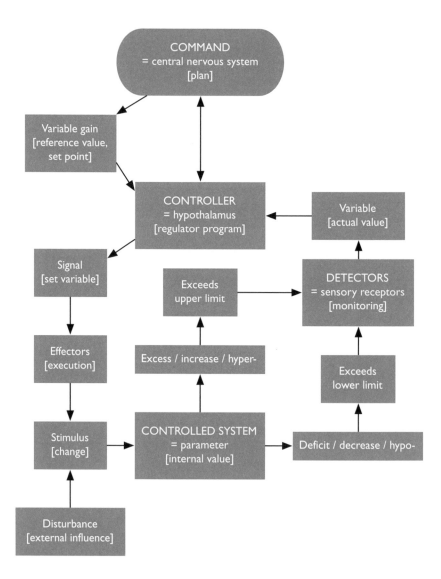

CHAPTER 1.

HOMEOSTASIS AND PHYSIOLOGICAL MECHANISMS

Physiology is the study of the function of living things. Human physiology is therefore the study of how we eat, breathe, move, think, have sex and do everything else we do. Classically, human physiology has been preoccupied with how our organs work; why the cells of every tissue are different and how these differences relate to the different tasks they perform. Superseding this however is one striking similarity, the pervading influence of homeostasis. With respect to this, it is most important to identify *the* overriding factors associated with underpinning life and then to understand the control of these factors. Furthermore, in order to study these mechanisms more effectively there is a need for an integrative and more even-handed approach to physiological measurement.

HOMEOSTASIS:

Homeostasis is the maintenance of stable conditions within a body despite changes in the external environment. Homeostasis depends upon the control and transmission of information. For complex organisms many activities need to be individually *regulated*. Regulation depends on the adoption of a set of rules that govern the adjustment of a process so that the process operates correctly. This adjustment requires a two-way exchange of information between a control centre and the controlled variable. The information is 'fed back' to enable adjustment of the necessary 'commands'. Feedback is a mechanism whereby the products of a process act as regulators of that process. In other words, the products regulate their own production. This feedback can be either positive and reinforcing, or negative and counteracting. Positive feedback self-perpetuates a process until subjected to negative feedback. Negative feedback opposes change in a given direction, but has a tendency to overcompensate (rebound). This then requires negative feedback in the other direction to rectify the imbalance. Two complementary but opposing negative feedback mechanisms therefore result in an ever-declining cycle of increases and decreases, until the optimum level of activity is achieved. It is this flexibility of the control system that allows for a limited degree of adaptation to changing environmental conditions and for a

6

For written notes:

7

harmonious integration of complementary internal processes. In short there needs to be room for adjustment, in order to maintain stable reactive conditions for metabolism independently of fluctuations in the environment. The control system can therefore be thought of as consisting of a regulatory operation with feedback. All parts of the body work by maintaining differences with both the internal and external milieu. This produces gradients that are maintained by controlling the rates of flow between the inside and the outside. Distinctions can be drawn between instances where one structure exclusively controls certain rates, and instances where different structures control the same rate but to varying extents. Many rates are controlled by many structures in order to fulfil specific functions. For survival, life depends on specific functions that facilitate the achievement of certain needs. These needs or wants are the characteristics of life:

1. Growth and size
2. Nutrition and feeding
3. Metabolism (material conversion and energy production)
4. Excretion and the elimination of waste
5. Motility (external movement for locomotion and internal movement for transportation)
6. Sex and reproduction
7. Defence and immunity

There is an additional characteristic common to all life that is not necessarily wanted, but it is essential, as life must pass through time:

8. Ageing and degeneration

The working of the body to achieve these ends is classically divided according to the systems used. These systems either refer to the separation of the body into different organs, or tissue types:

Organs
1. Cardiovascular
2. Respiratory
3. Gastro-intestinal
4. Endocrine
5. Nervous
6. Renal (urinary)

For written notes:

9

7. Musculoskeletal (locomotive)
8. Reproductive

Tissue types
1. Structural (bone and cartilage)
2. Muscle (cardiac, skeletal and smooth)
3. Nerve
4. Lining and covering (epithelial membranes)
5. Secretory (glandular)
6. Binding and packing (connective; e.g. ligaments, adipose)

There are many obvious parallels between these groupings. For example, external movement depends heavily on the skeletal system, which in turn is served by bone and muscle tissue. Sometimes the division is not clear, such as with skin (integument), as it can be considered to be either an organ or tissue type. Furthermore, the immune system does not have any clearly defined organ. Neither do blood or lymph fit into any clear tissue type (although they are frequently classified as connective tissue). However, there is an alternative classification. The characteristics of life may use certain organ systems and tissue types, but the ultimate aim is to regulate certain factors. How does a body stay in balance, remaining alive and distinct from the surrounding environment? Any living body needs to maintain certain factors within normal physiological ranges. This is achieved by many mechanisms employing a wide range of filters and gradients. Principally, differences in concentration across barriers are controlled, by selectively allowing the passage of certain things between one side and the other. Regulation of any factor in this way depends on feedback in order to maintain an optimum level. It is the limitations of the feedback processes that lead to deficiencies or surpluses, and ultimately clinical consequences. In order to achieve the essential characteristics of life, the body as a whole is dependent on the balance of certain interrelated key factors:

1. Volume
2. Hydration
3. Oxygenation
4. Energy
5. Acidity
6. Temperature

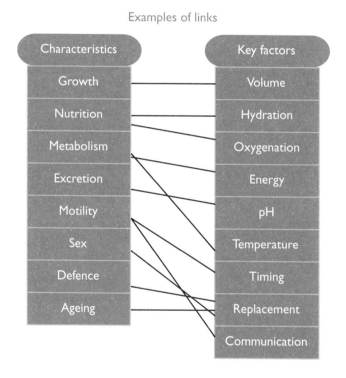

Examples of links

7. Timing
8. Replacement
9. Communication

Volume distinguishes the space that is occupied by life. Hence, the control of volume is what separates what is alive from what is not. Volume needs to be maintained with respect to the boundaries and pressures between things. Everywhere throughout a living body boundaries are acting as filters, and pressure provides a gradient to varying degrees and in different directions. Water is a critical determinant in producing pressure and providing a means for the expression of gradients.

Hydration to varying degrees will affect all other processes as it is water that provides the basis for life. Water is the medium on and through which everything in life depends. At the physical level the interaction between liquids and gases is vital.

Oxygenation and gaseous exchange provide a fundamental role in the chemical production of energy. Either aerobic or anaerobic respiration involving gaseous exchange is a basic need of all life.

Energy is the means by which all other ends are met. In life, energy is used to produce order. The use of energy is what makes processes active and distinguishes life from death.

Acidity and temperature must be regulated in order to maintain a stable and predictable reactive internal environment. Amongst other things, this is necessary for the efficient use of energy.

Timing is crucial for the relay of information and execution of commands. The balance between supply and demand must be carefully regulated and timed to prevent deficit or surplus. Within the body timing is necessary for everything, from the rate of nervous signal propagation to the transfer of intracellular messengers. Furthermore, to efficiently connect the body with the outside, timing is needed to entrain internal biological rhythms to external environmental cycles.

The cells responsible for any process can be - and frequently are - subjected to attack and damage with the passage of time.

FOR WRITTEN NOTES:

13

Replacement depends on the balance between construction, destruction and repair. In particular, defence and regeneration rely on metabolism and cell division to counter degeneration and loss. This maintains the building blocks of the main organ systems, which in turn facilitate the execution of all the active processes necessary for life.

Communication dictates all the other processes, in order to maintain the whole body in balance. This is achieved by providing control signals in response to feedback, and whenever possible by linking common feedback information between different processes. Thus enabling the simultaneous integration of all the key factors necessary for life.

Indeed we should consider that these factors that require regulating are themselves the basic needs of life. There is however more to it than that. It is only by integrating all of the active processes involved that the essential characteristics of life can be achieved. The balance of all these key factors *together* is absolutely vital for the life of a body. The solution is provided by how all these key factors are linked together in an interdependent way. This complex puzzle of inter-related factors is the mystery of life that we seek to understand. To this end, we can adopt a mechanistic approach.

MECHANISMS:

In physiology a mechanism is considered to be a system of mutually adapted parts working together. All function depends on the evolution of corresponding structure. This is clear from Darwin's law of evolution by natural selection. The development of structures requires energy, and the efficiency of energy expenditure by a body will be correlated with the probability of that body surviving. Therefore in order to understand the underlying cause of any particular function we must think of a relevant evolutionary pressure, which would lead to that function enhancing survival. On a practical level, it is necessary to make appropriate measurements to achieve a balanced view of any physiological mechanism. In particular, establishing interrelationships in cause and effect refines the understanding of any such system. When considering causality it is necessary to identify certain essential criteria. There are five fundamental factors that I propose must be studied in order to establish the full scope of any measurable effect:

ANECDOTES

When making measurements:

Do not change more than one variable at a time.

For evaluation ensure you are comparing like with like.

If a measure is indirect, check to what proportion it is dependent.

A correlation can be confounded by hysteresis.
Hysteresis is a difference in outcome
when comparing increasing with decreasing
the variable being measured.

Ensure averages are representative of the numbers being summarized:
Mean = sum / sample size
Median = middle
Mode = most frequent

The average spread or dispersion from any true value is approximately
proportional to the square root of the number of measurements.
One standard deviation (SD) is the square root of the average of
the squared deviations of the observations from their mean.
This deviation as a proportion of the true value
therefore decreases as the sample size increases.

1 SD = square root of [sum of differences (between each value and the mean)^2
/ (sample size -1)]

Coefficient of variation = SD as % of mean.

1. Correlation
2. Resolution
3. Adaptation
4. Limitation
5. Reproduction

Correlation:

Correlation is the association (or dissociation) between two variable measurements. This is sometimes expressed by a correlation coefficient (r, sample correlation). The coefficient can range from a perfect positive correlation (r = 1.0), to a perfect negative correlation (r = -1.0), the mid-point (r = 0) indicates no relationship between the two variables. For example, there is a positive correlation between height and weight, as an increase in one of these variables tends to coincide with an increase in the other. Ideal body weight (in kg) is derived approximately by the Broca index, which is equal to height in centimetres - 100. The standard average body weight of 70 kg is therefore 1.7 m tall. The actual body weight is usually expressed as the body mass index (BMI or Quetelet index), which is the weight in kilograms divided by height in metres-squared. An example of a negative correlation is age and mental agility, as increasing age tends to coincide with a decreasing mental agility. The coefficient of determination (r^2, squared correlation) is the square of r and represents the proportion of variance in one variable that can be accounted for by the second variable. On the other hand, the regression coefficient (b) is the slope of the line in a simple linear regression:

$$Y = a + b.X$$

Where; Y = dependent variable (effect, criterion, outcome, response or predicted variable),
a = intercept (the predicted value of Y when X is zero),
b = slope (the amount of change in Y for a change of 1 in X), and
X = independent variable (cause, explanatory, or predictor variable).

This form of 'co-relation' between two variable measurements makes several assumptions that are common to many ideal statistical calculations. Firstly, the data should have normal (Gaussian or parametric) distributions. This means that all the numbers for X and Y plotted separately must be symmetrical around their respective mean values, with 95 % of the points falling within + and - 1.96 standard

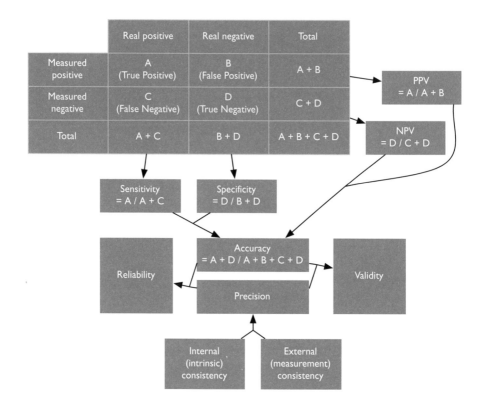

Specificity = True Negatives / (True Negatives + False Positives)

Negative Predictive Value (NPV) = True Negatives / (True Negatives + False Negatives)

Sensitivity = True Positives / (True Positives + False Negatives)

Positive Predictive Value (PPV) = True Positives / (True Positives + False Positives)

Thus specificity excludes false positives, whereas sensitivity excludes false negatives.

deviations, without significant skew or kurtosis. Secondly, the data should have homogeneous variance (equal scatter, homoscedastic or homoskedastic). This means that the variation between X and Y should not be significantly different (i.e. standard deviations should be equal). An F-test can be used to compare the variances of two groups, this is in contrast to a T-test which can be used to compare the averages of two groups; the statistical significance of the two tests can vary independently. Finally, successive measurements should be independent of each other, meaning that the data represent a random sample from the population of interest. True randomization should itself also ensure that the data collected is representative of the population.

These are not necessarily cause and effect relationships as other variables may result in a higher correlation. Cause and effect may depend on the culmination of several slight correlations between many variables. Indeed, opposing correlations may conceal the true variable causing an effect. A correlation must exist either spatially or temporally. That is to say that there will be one or more correlations in both space and time with respect to the measurement being made. The measurement will be made with reference to a particular *site* and at a particular *time*. This measurement will therefore vary according to any changes in that site or time. Take the example of a dose-response relationship. The biological gradient that is measured will have a variable effect coincident with differences in the site and time of both exposure *and* measurement.

Resolution:
Resolution is the minimum difference between two measurements at which they can be seen as separate rather than as the same. (This is distinct from discrimination, which is the separation of a single entity.) The degree of resolution is crucial to the interpretation of a measurement. Resolution can also be called selectivity, and is divided into specificity and sensitivity. Both specificity and sensitivity relate to the variability of a measurement. However, there is much confusion over the terminology employed.

Specificity refers to how close a measurement is to the actual true value. Any true value is by definition unknown, and is often taken as the degree of correlation of the measuring instrument with an established 'gold' standard instrument for calibration. In order to ensure there has

19

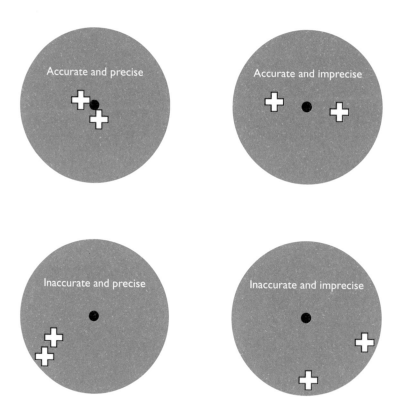

been no drift in the known reference values and consequent change in specificity such calibration should be performed both before and after making measurements. Alternatively a specific test is often referred to as having a high true negative rate, corresponding to a low false positive rate. Thus a specific test correctly identifies both true negatives and false positives (i.e. in effect identifying all negatives). To take this further, the negative predictive value (NPV) is the probability that a negative is truly negative. A specificity exceeding 1 in 1,000 is required for a law of nature. In general parlance, specificity, validity and accuracy all mean the same. A systematic error (a consistent mistake when measuring) will limit accuracy, as measurements so affected, are always either too small or too large.

Quite distinctly, sensitivity refers to how fine the differences are between repeated measurements. A random error (an inconsistent mistake) will limit sensitivity, as measurements are equally likely to too small as too large. In this case, sensitivity, reliability and precision can be considered interchangeable terms. Alternatively a sensitive test is often referred to as having a high true positive rate, corresponding to a low false negative rate. Thus a sensitive test correctly identifies both true positives and false negatives (i.e. in effect identifying all positives). To take this further, the positive predictive value (PPV) is the probability that a positive is truly positive. It is not true that increasing sensitivity simply enables the detection of more 'noise'. With sensitivity it is important to consider the level of detail appropriate to the context. A measurement can be too sensitive to answer the prescribed question. One common analogy of this is that when you are in amongst the trees it is impossible for you to see the picture of the whole forest.

Sensitivity must be considered in addition to considering the level of specificity appropriate to the stage of knowledge. Start with what is known, and work towards what is unknown with increasing detail. Using our earlier analogy of the picture, you should be certain of what the picture is about, before you try to find out what the paint is made from. In either case, the frame of reference depends on the degree of experience. To confound things further, it is also possible to think of resolution in terms of generalities and specifics. For example, differences in the level of magnification may be referred to as systemic or local phenomena. Systemic phenomena have less magnification and so are general, whereas local phenomena have more magnification so

Abbreviations for Orders of Magnitude in Measurement:

Name	Unit Prefix	Equivalent Multiplier	Suffix
Yocto	y	10^{-24}	septillionth
Zepto	z	10^{-21}	sextillionth
Atto	a	10^{-18}	quintillionth
Femto	f	10^{-15}	quadrillionth
Pico	p	10^{-12}	trillionth
Nano	n	10^{-9}	billionth
Micro	μ	10^{-6}	millionth
Milli	m	10^{-3}	thousanth
Centi	c	10^{-2}	hundreth
Deci	d	10^{-1}	one-tenth
Deca	da	10^{1}	ten
Hecto	h	10^{2}	hundred
Kilo	k	10^{3}	thousand
Mega	M	10^{6}	million
Giga	G	10^{9}	billion
Tera	T	10^{12}	trillion
Peta	P	10^{15}	quadrillion
Exa	E	10^{18}	quintillion
Zetta	Z	10^{21}	sextillion
Yotta	Y	10^{24}	septillion

they are specific. This confusion is unlike typical scientific terminology. Every word usually has a distinctly different meaning. However, as long as there is neither ambiguity nor misunderstanding, the *precise* term is unimportant as long as it is *accurate*.

Adaptation:
The measurement of any active process is subject to dynamic change due to adaptation. Adaptation is an alteration in the degree (gain) of a response due to a difference in the timing of stimulation, or indeed measurement itself. In other words, a different reaction is measured with exposure to a change in conditions. This may be due to a change in the interval between successive recordings, or simply due to recordings being repeated at regular intervals. Adaptation occurs because of differences in magnitude or velocity between measurements. In other words, the proportion of the maximum and the difference in the velocity of the factor being measured will affect the measurement. When considering proportionality, it is useful to think of changes in magnitude, level or intensity (where one order of magnitude is one ten-fold difference in size). For differentiability, think of changes in velocity, frequency or rate. The concept is the same whether we are talking about the amplitude and frequency of sound waves, the proportionality and differentiability of sensory receptors, or the intensity and rate of epidemiology cases. Take the epidemiological example of prevalence and incidence. Epidemiology describes disease pattern, identifies contributing factors, and provides management data. Within this context, prevalence is the number of existing cases, whereas incidence is the number of new cases occurring; both within a known population size and at a given time. Prevalence is obviously not the expression of a rate, whereas incidence is. To confuse matters, both of these should further be expressed as a proportion of the cases found in normal individuals (control-adjusted). This has particular relevance for the calculation of risk when exposed to a hazard. Risk may be relative, attributable or absolute. Relative risk is the ratio of exposed to non-exposed incidence rates. Attributable risk is the difference between exposed and non-exposed incidence rates. Whereas, absolute risk is the exposure incidence rate divided by the dose-response. It is clear that care must be exercised when interpreting these terms. The expression of both *intensity* and *rate* depend on the relative differences between measurements.

For written notes:

23

Limitation:

There are of course limits to every measurement. Limitation is the restriction of a variable between absolute minimum and maximum values, the range of which reflects the stability of the parameter being measured. The process is characterised by both the presence of thresholds and reversibility. The *scale* of a measurement encompasses the normal physiological range; typically represented by accepted clinical guidelines of minimum and maximum threshold values. A scale may be nominal, ordinal, or numerical. A nominal scale is expressed as a purely qualitative category, for example gender. Whereas, an ordinal scale is expressed as a quantified category, for example grade. This is in contrast to a numerical scale which is expressed as a continuous quantity (either interval or ratio), for example blood pressure. The size and equality of intervals between measures, and the reference 'zero' or 'set point' value of one scale compared with another scale must be considered during interpretation. On the other hand, reversibility is an equally important concept to the limitation of a measurement. There is a point of *balance* beyond which a process may be driven in an opposite direction or have an opposite effect. Any given measurement may incorporate a change in polarity, or it may be different if made when the parameter is rising compared with when it is falling. It should also be remembered that without the negative the positive would not be recognised, and so the same emphasis should be attributed to both. Thus, contrast is needed in order to recognise any difference in state. When considering control, the concept of feedback and auto- or self-regulation relies heavily on these types of limitation.

Reproduction:

The reproduction of measurements is fundamental to scientific investigation. Reproduction is the production of copies, or the repeatability of measurements. There are two key underlying principles, that of possibility and probability. In the collection of evidence, the possibility (or feasibility) of a measurement being reproduced corresponds to there being a *plausible* theory of cause and effect. Usually, it is possible to relate to an extant basis of coherent theory, or to an experimental analogy. There are typically a number of theoretical paths that may be true. Pathways can be alternative (instead of), for support (as well as) and for contingency (in case of, or auxiliary). When considering plausibility, be cautious of a single piece of evidence masquerading as a consensus of opinion. A disproportionate emphasis

ANECDOTES

Seeing observes the unexpected,
looking eventually finds what you expect.

Be aware that overemphasis of significant results
at the expense of non-significant results
causes an over-estimation of effects.

Care should be taken when making extrapolations and inferences.
A little knowledge can be a dangerous thing!

Ensure that the solution is not more damaging than the problem.

Everyone makes mistakes.
Making errors is the most effective way to learn,
but only if there is an opportunity for redress.

is placed on the so-called "facts" that are propagated through generations of literature, without a full comprehension of their source.

Similar to plausibility, the probability (or confidence) of reproduction is crucial to our understanding of making a measurement. Probability impinges on the context of a measurement with respect to other measurements. This relates to the *significance* of the measurement itself, i.e. the likelihood of its occurrence and consistency of repetition. This significance can in turn be interpreted in terms of consequence. Both shortage and excess adversely affect efficiency and thus have consequences for survival. Values that are too low result in deficiencies, and values that are too high result in surpluses. In all cases, care must be taken when considering significance. All measurements are intrinsically equally important regardless of significance. It is not how many significant differences you find, but how many you find significant out of how many differences you look at. Nevertheless, individual significant measurements are often valued more highly than individual non-significant measurements. Moreover, finding many measurements of apparently no significance is itself significant. Unfortunately, the lack of emphasis on, and consequent underreporting of, such measurements results in a bias in favour of individual significant measurements. The most common limit used to determine statistical significance, or the conventional probability value (p-value), is 0.05 (representing 5 % of the population being studied having the characteristic concerned). Clearly a difference may be found to be statistically significant which is not biologically significant.

Power calculations:
The statistical significance (or probability) of the difference between two groups of measurements can also be thought of as discriminating power. Broadly speaking, discriminating power is the ability of one method of measurement to distinguish differences compared with another method of measurement. The number of repeat measurements needed to separate an effect can be calculated by analysis of the power of a study. Such power calculations are an essential consideration; not least to ensure that there is not economic wastage in collecting too much data, or that significant differences are not missed due to insufficient data.

The minimum number (n) of measurements needed will depend on the difference between the two mean values that is to be distinguished,

ANECDOTES

Rigorous scientific method should be first and foremost transparent.

Do what you say and say what you do.
Be honest, your word is who you are.

Science is the quest for truth.
Dogma delays the advancement of science.

The difference between believing something to be true
and knowing something to be true is a scientific argument.

the standard deviation of those means, and the acceptable level of probability for both true and false positives. This calculation is made based on the standard normal distribution (z-distribution), which has a mean of zero and standard deviation of one.

$$n = [(z_\alpha + z_\beta).(\sigma)/(\mu_1 - \mu_2)]^2$$

Where; z_α = two-tailed z-value of a false positive error (+/-1.96 for 95 % of the area under the curve; equating an upper and lower 2.5 % probability of a false positive), z_β = lower one-tailed z-value of a false negative error (-1.28 for 10 % of the area in one tail; equating a 90 % probability of a true difference, either positive or negative), σ = population standard deviation, μ_1 = first population mean, and μ_2 = second population mean. For prospective power calculations the values for these parameters are all chosen based on the best available evidence.

It looks jolly simple doesn't it? This complicated formula is central to understanding the context of all measurements. Fortunately there is a much simpler equation that can be adopted in practice. A rough and ready rule of thumb for a study with a P-value of 0.05 is the squared ratio: $(\sigma/(\mu_1-\mu_2))^2$. A sample size twenty-times this squared ratio is required for both groups if a 90 % probability of detecting a true difference is desired. A multiplier of fifteen-times can be applied if only 80 % is adequate.

If a prospective power calculation has not been made, then a retrospective analysis should still be applied in order to evaluate both the power of the study and the magnitude of the difference that could have been detected as statistically significant. Remember, absence of evidence is not evidence of absence.

Statistical anomalies:
To complicate matters, there are several possible implications when a difference is looked for. A true difference may be found, called a true positive. A bogus difference may be found, called a false positive (type I, or α error). Similarly, a true difference may be missed, called a false negative (type II, or β error). Alternatively, there may be no difference found when there truly is no difference, called a true negative. Strictly speaking, the four possible outcomes above would occur with

ANECDOTES

29

Statistics strives to derive order from disorder,
whereas dynamics derives one form of order from another form of order.

When planning research, it is important to distinguish between an objective and an aim:
An objective is a goal that can be achieved by the proposed task.
An aim is an ultimate goal that cannot be completely achieved by the proposed task.
Thus an aim is a reason for a study, while an objective forms the basis of a hypothesis.

Occam's (Ockham's) Razor:
When making a choice between hypotheses,
start with the hypothesis that makes the fewest assumptions.
Or rather, always choose the simplest.

equal probability, if there were no experimental prejudice. However, experimental measurements are generally designed with the intention of exposing differences. Therefore there is a bias in favour of finding such differences. If one looks hard enough for a difference, one will find that difference. In this way, determination can find justification for any theory. Inaccurate theories can be reinforced and perpetuated. It takes integrity to reject a statistically significant difference that is due to a chance coincidence, especially when it provides evidence supporting an accepted theory. When under pressure to succeed, this is exploited by the individual researcher in the blind desire to produce 'positive' results. This taints the whole evidence base of scientific publication. The problem is further exacerbated by the corresponding under-reporting of 'negative' results. This obviously handicaps the progress of the scientific community as a whole. Challenges to accepted science always increase understanding and provide evidence to strengthen theories. There is one cruel exception to this rule, which is ultimately counterproductive to the advancement of science. That is the intrinsic predisposition in favour of differences. To avoid such a bias, both measurements (whether laboratory, clinical or epidemiological) and statistical analysis should follow a carefully planned protocol to objectively test a predetermined hypothesis. A null hypothesis states that there is a significant difference between X and Y. If untrue, then one can reject the null hypothesis and accept the alternate hypothesis, that there is a significant difference between X and Y. However, there is still the risk of selective attention bias. It seems likely that many investigators find differences that do not exist due to the inadequate use of controls. Both negative and positive controls are needed to ensure that a real effect is measured, and not a false negative or false positive respectively. A control repeats exactly with the exception of the intervention alone. It is difficult to ensure that a control is an exact replicate. In any case, repetition itself implies occurrence at a different time, and this in itself reveals a difference.

In addition, the best way of not becoming too involved in statistical subtleties is to choose a problem for which a large difference is likely to be found on plausible grounds, to ensure that a large sample of data can be collected, and that the method of measurement is as precise as possible. Statistical analysis will then be comparatively simple and should provide an unequivocal answer. Be cautious of excessive use of statistical analysis. With judicious use of statistics anything can

31

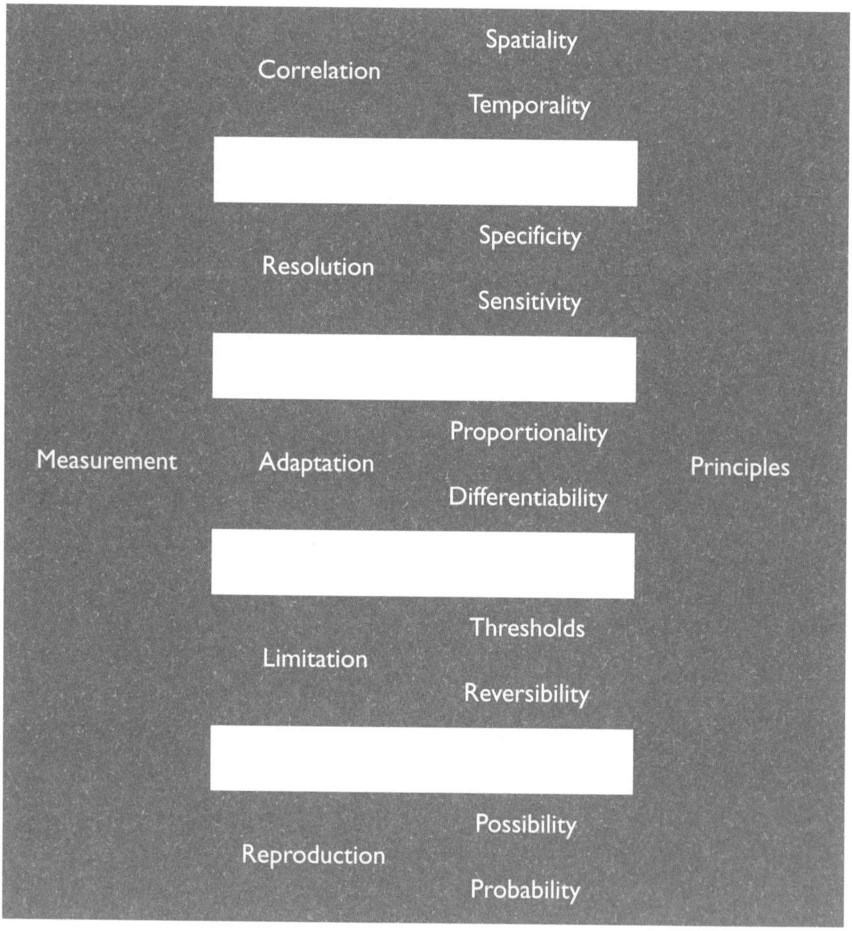
Correlation
Spatiality
Temporality
Resolution
Specificity
Sensitivity
Measurement
Adaptation
Proportionality
Differentiability
Principles
Limitation
Thresholds
Reversibility
Reproduction
Possibility
Probability

be proven. Indeed, a declaration aspiring to scientific proof should always arouse suspicion. Philosophically, it is only possible to *disprove* and then only with a reasonable degree of certainty (usually 95 % probability). The only accepted facts are those theories that have stood the many tests of time, represent a consensus of opinion, and have been elevated and enshrined as scientific laws; usually a century or more after the original advocators have met their demise.

SUMMARY OF MECHANISMS:

All of these factors need to be considered to fully understand a physiological mechanism. Indeed, this theory on measurement could - perhaps should - be applied to all practical science. Each of the five fundamental factors can be further divided into two, to give rise to a total of ten different principles:

1. Spatiality
2. Temporality
3. Specificity
4. Sensitivity
5. Proportionality
6. Differentiability
7. Thresholds
8. Reversibility
9. Possibility
10. Probability

This includes every conceivable aspect that can change in any measurement. In general though, measurements are rarely - if ever - expressed in the full context of all these aspects. Nevertheless, together they represent the causality criteria. Exhausting all these principles should yield a complete picture for any phenomena. Each of these principles should be considered separately and attributed equal weight. It is hoped that a greater awareness of the principles will help scientists in all fields recognise new areas for focus.

Science reveals an understanding of the working of an object or process by logical deduction. Furthermore, it is the *belief* of scientists that everything has a meaning; and that meanings can be understood by the application of scientific measurements. Quantification is the essence

ANECDOTES

Every answer is just a question of how much detail.

If the explanation is complicated,
it is because the person explaining either does not understand it,
or they are trying to confuse you.

Don't make anything more complex than it already is.

33

of science. It follows that careful observation of the differences between measurements is the primary tool of the scientist. Thus anything that highlights such differences is welcomed. In this way, the ten principles above should augment traditional brainstorming (mind mapping) on measurement. Of course the questions reflected on should still include: *what, where, when,* - and most importantly - *why* and *how*. Also remember that at the most basic and simple level all measurements affect themselves. The application of a measuring instrument of any type will alter the value measured. Depending on the underlying context there may also be a level-dependent systematic change in the measurement caused by what is being measured (i.e. the magnitude of a parameter can affect the properties of that parameter). Critical thought should always be applied to identify potential confounding factors (e.g. measurement artefacts). First of all, whenever a measurement is taken check the order of magnitude. Any anomalous values should be traced to their source. Secondly, whatever the method of measurement it is important to consider in what way would the response have been different if there had been no such intervening measurement. It follows that a change in any one of the ten principles described above will cause a change in all of the other nine principles. Conversely, also bear in mind that some of the most profound scientific discoveries relate to not being able to measure any change at all (e.g. the constancy of the speed of light). Finally, measure what is important. Don't consider everything that can be measured to be important.

34

For written notes:

35

CHAPTER 2.

VOLUME

INTRODUCTION TO VOLUME:

Volume is a measure of the magnitude of three-dimensional space enclosed within or occupied by a body and as such is a representation of its quantity. All matter (living or otherwise) is anything that has both mass and volume (rf. density = mass x volume). Here volume is emphasized as a measure of the physical presence of an organism and also as an important factor to the active processes essential for life. Volume depends intrinsically upon both pressure and the integrity of boundaries. Hence, the forces exerted across membranes are crucial factors for consideration. The processes that particularly emphasize this importance on a macro scale are: growth, nutrition, excretion, and transport (of gases, liquids and solids). The systems that primarily need control are the cardiovascular, gastro-intestinal and renal systems. However, the control of overall body volume and shape has more obscure origins.

THE MAKING OF DIFFERENT BODY SHAPES:

We all know that the structure and function of living things (organisms) depends upon their construction from cells, tissues and organs. Organisms are comprised of organs (sometimes called organ systems), and organs are comprised of tissues, which are in turn comprised of cells. We also know that there are certain key factors that are necessary for life to be maintained (volume, hydration, oxygenation, energy, acidity, temperature, timing, replacement, and communication). These collections of cells, tissues and organs facilitate the same key factors in all organisms. However these organisms are vastly different. Consider the diversity of shape: from scorpion to squid, from shark to seagull, or even from microbe to 'man'! Further consider the diversity of habitats that living shapes have evolved to inhabit: from desert to ocean, from tropical rainforest to Artic ice-flow, or even from city to space station!

In order to investigate body shape further it is necessary to employ taxonomy. The study of taxonomy is the classification of organisms into a hierarchy of groups based on the similarity of their characteristics.

VOLUME

FOR WRITTEN NOTES:

37

With respect to this, the smallest indivisible category of organisms is that of the species. Only members of the same species may produce offspring that are themselves capable of reproduction. Organisms that have sex which produces offspring that are incapable of reproduction (sterile) are chimaera. Examples include: the liger (tigron) as the offspring from a lion and tiger (which are from different continents); the mule from a female horse and male ass; and the hinney from a male horse and a female ass (NB a donkey is a domesticated ass). By definition the genes of chimaera will not be carried through subsequent generations or contribute to the evolution of new body forms.

All bodies, whatever their shape, are made from cells. The cell is the smallest indivisible category of life. Hence, cell theory espouses first of all, that all free-living organisms are composed of cells and their products. Secondly, all cells are basically similar in their chemical construction. Thirdly, that new cells are formed from pre-existing cells by cell division. And finally, that the activity of an organism is the sum of activities and interactions of its cells. A cell comprises a membrane bound aqueous solution, capable of creating copies of itself by division. The principal components of a cell are, the: nucleus (containing deoxyribonucleic acid), cytoplasm (the cellular medium), mitochondria (for ATP synthesis, and can be up to 20 % of cell volume), Golgi apparatus (for vesicle secretion), endoplasmic reticulum (for protein manufacturing), lysosomes (which engulf organelles for recycling) and cytoskeleton (for shape and movement). The cellular 'unit' has its limitations. Cell size is approximately constant in all tissues and organisms, between about 2 μm and 200 μm. Cell size is limited by the physical restraint of surface area to volume ratio decreasing as size increases. Critical size depends on membrane integrity and internal transport mechanisms, particularly the limits of diffusion.

Differences in cellular demand can result in a need to specialize. Specialization occurs in order to exploit and colonize extreme physical environments and compete for ecological niches, and results in increasing complexity of physical adaptations. Specialization correlates positively with dependence and hence also with increasing vulnerability. It is the increasing need for specialization that leads from single cellular to multiple cellular (multicellular) organisms. Eventually, over time, extreme levels of complexity are reached. For example, the total number of cells in the human body is about 75 x 10^{12}. One third of

VOLUME

FOR WRITTEN NOTES:

39

these cells (25×10^{12}) are erythrocytes (red blood cells or corpuscles) the most common cell type, each typically $7 \mu m$ in diameter; and 25×10^9 are neurons (not neurones) which group together to form nerves, the longest being the sciatic nerve which conducts signals up to 1 m between the spine and the foot.

With increasing specialization, cells form tissues and tissues form organs. A tissue is a group of cells embedded in extracellular matrix and having similar structure and function, e.g. bone, muscle, glandular, nervous, epithelial or connective tissue. An organ is any group of tissues that performs a specific role, e.g. heart, lung, liver, or kidney. These organs in turn give rise to different systems, e.g. cardiovascular, respiratory, gastro-intestinal, endocrine, metabolic, nervous, urinary, musculoskeletal, or reproductive system. All of these systems are combined under the umbrella of a single body plan, which terrestrially favours cephalization, pentadactyl limbs and an upright posture. The size possible by using this interdependent arrangement of organs ranges from a few grams to many tonnes.

In a sense, life is an increase in diversification and a reduction in randomness of the environment. A non-uniform environment provides niches for life to colonize. It is the variety of habitats that provide the differences that are necessary for the evolution of a multi-layered diversity of life. Different habitats can support a varying number of trophic levels. There may therefore be different numbers of steps between the primary provider and the topmost predator. Fewer steps mean fewer species. Fewer species means more energy is available for each species. Hence there will be either larger organisms or larger numbers of organisms of each species. Thus, low species diversity generally coincides with high species productivity and vice versa. Overall, to maintain a constant species diversity old species must become extinct at the same rate as new species are generated by evolution.

An example of a low species diversity habitat is that of the cold Antarctic Ocean with its larger numbers and larger size of algae, krill and whales. Here both low temperatures and the need to survive protracted winters without feeding requires a larger body size. The larger size is not only to decrease the efficiency of heat loss and so promote heat retention, but also as a strategy to survive periods of nutrient deprivation when there

Volume

For written notes:

41

is no sunlight. Colder water also dissolves more oxygen and so is more productive when there is sunlight so growth can be more rapid.

In comparison, an example of a high species diversity habitat is that of a tropical coral reef with its dazzling display of different organisms. Closer to the equator there is a diurnal cycle of light, a smaller tidal range and comparatively continuous warmth. This less dynamic physical environment enables more energy to be diverted to the exploitation of smaller differences between niches than would otherwise be possible. This of course results in a larger number of more similar species. Within all habitats, there are very similar niches that have only small physical and chemical differences between them. However, in the tropics there is more energy available to exploit small differences. It follows that habitats with high species diversity will by necessity have a much higher degree of inter-species (between different species) competition. Compare this with habitats of low species diversity, where there is a higher degree of intra-species (within the same species) co-operation. In the example of the Antarctic, organisms from the same species act together: migrating, shoaling, or hunting as a group. The more unique or hostile an environment, then fewer species are able to exploit it, the less inter-species competition there will be, so populations can grow larger and co-operate to improve survival. On the coral reef there is much more individuality, although there are still examples of substantial co-operation (manifest in examples of symbiosis; i.e. where two species are intimately inextricably interdependent). Co-operation confers a greater advantage than competition, but only when there are sufficient resources to satisfy all those who are co-operating.

There are of course other factors to consider that affect diversity. For example the availability of a stable surface increases the scope for niches, by providing a fixed reference point for interfacing with other species. Generally speaking there are more species where there are surfaces (at the interface between two physical states), consequently organisms tend to be smaller, e.g. consider algae at the surface of an ocean, or a microbial film on the surface of a whale. This also highlights the fact that the larger the organism and more diverse the shape, the more potential niches its body can provide for colonization. The generation of new surfaces (or the cleansing of old surfaces) provides room for expansion. It follows that the colonization of virgin territory increases the opportunity for evolution. Thus, a changing surface

FOR WRITTEN NOTES:

43

(e.g. sloughing of skin, plate tectonics) clears the way for recolonization by subsequent generations, and so drives adaptation by competition for the 'new' surface.

PHYSICS OF VOLUME AND SURFACE TENSION:

Volume is measured in litres (l). One litre being equal to one cubic decimetre (1 l = 1 dm³). This differs from the volume occupied by one kilogram of water (at its maximum density and standard atmospheric pressure) by twenty-eight parts in one million. Thus, volume should strictly speaking be expressed in cubic metres, not litres, especially when a high degree of precision is required. The metre (m) is the basic unit of length, where one metre is equal to 1,650,763.73 wavelengths of radiation corresponding to the transition between the electron orbital energy levels $2p_{10}$ and $5d_5$ of the krypton-86 atom in a vacuum. Incidentally, the unit of mass is the kilogram (kg), and is equal to the mass of the international standard made of platinum-iridium. Derived from this is the unit of pressure, the Pascal (Pa), expressible as $kg.m^{-1}.s^{-2}$, or alternatively as $N.m^{-2}$. However, measures of blood pressure are still referenced to mercury (Hg) sphygmomanometers, where 1 mmHg (1 torr) is equivalent to 133.32 Pa (or 13.6 mmH_2O). Hence, the standard atmosphere (1 atm; the air pressure at mean sea level and 15 °C) is frequently expressed as either 101.325 kPa, 101.325 $kN.m^{-2}$ or 760 mmHg (dry; that is to say with zero water vapour pressure or 0 % relative humidity). Surface tension is expressed in units of $N.m^{-1}$ or $kg.s^{-2}$.

At the interface between gases and liquids surface tension is obviously important. Interactions between molecules at the surface of a liquid make the surface resist stretching. The higher the surface tension the harder it is to stretch. In the lung, detergents reduce surface tension by disrupting interactions between surface molecules. The lung has a mixture of detergents called *surfactant* produced by Type II alveolar cells (Type II pneumocytes). Surfactant is a surface-active substance forming a phospholipid film covering the inside of the lungs. Surfactant reduces surface tension when the lungs are deflated, but the surfactant spreads more thinly as the lungs inflate. Therefore the surface tension is very high at maximal inhalation. So little breaths are easy, where the contribution of surfactant to reducing surface tension is greater, but big breaths are hard.

44

45

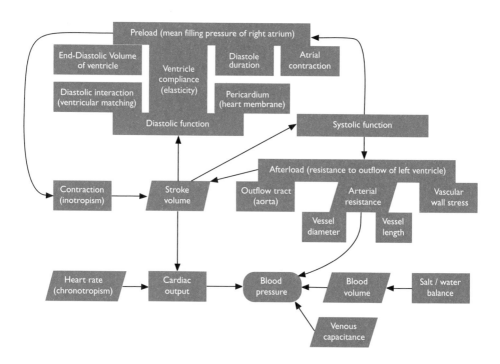

Bubbles are formed when a film of fluid surrounds gas, the film shrinks, compressing the gas until eventually there is equilibrium between tension and pressure. Big bubbles have low pressure and little bubbles have high pressure. It follows that if a big bubble is connected to a little bubble air will flow from high pressure to low and the little bubble collapses into the big bubble. Hence, big bubbles 'eat' little bubbles. In the lung, alveoli form an interconnecting set of bubbles. The tension and pressure of bubbles can be linked by Laplace's law:

Pressure = 2 x surface tension / radius.

If Laplace's law is applied to lung expansion, big alveoli would 'eat' little ones, and the lungs become a physical impossibility. However, as alveoli get bigger surface tension in their walls increases because surfactant is less effective, so pressure stays high and stops them 'eating' little alveoli. In respiratory distress syndrome, babies born prematurely, do not have enough surfactant. Their lungs are very stiff, with fewer, large alveoli; so breathing and gas exchange are compromised.

CARDIOVASCULAR RESPONSES:

Blood is the principal intermediary for water transport, and primarily in conjunction with the kidneys, is responsible for the maintenance of body volume. At various points in the circulation there are various receptors and effectors. The medulla oblongata (vasomotor centre) co-ordinates this nervous regulation of the circulation. The lateral part of the vasomotor centre elicits *pressor* responses (increasing blood pressure); whereas, the mediocaudal parts elicit *depressor* responses (decreasing blood pressure). In addition, there is hormonal regulation of water volume by the hypothalamus.

The four main factors that require control are the: total peripheral resistance (systemic vascular resistance), cardiac output, vessel capacity and blood volume. The total peripheral resistance of the arterioles and arteries (resistance vessels) and cardiac output are both adjusted to maintain the blood-pressure gradient that is necessary for blood to flow around the vascular system. An increase in one of these is compensated for by a corresponding decrease in the other, and vice versa. Similarly, vessel capacity and blood volume are adjusted to maintain a static blood pressure. Changes in vessel capacity depend

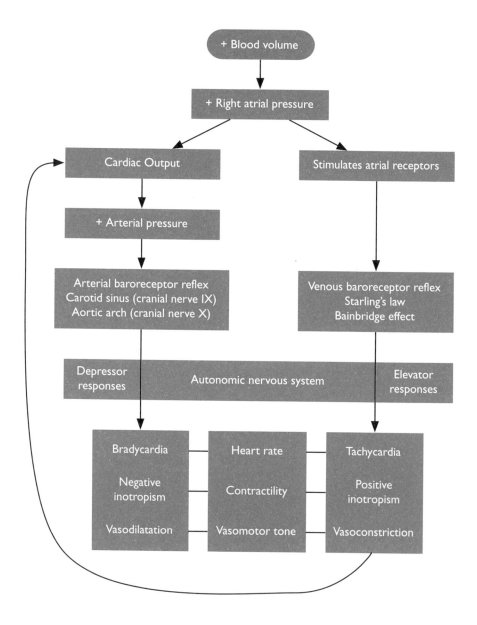

predominantly on the vasomotor state of the veins (capacitance vessels). Whereas, changes in blood volume depend on the filtration and reabsorption in both capillaries and kidneys. Maintenance of circulating blood volume is crucial. A reduction in intravascular volume causes a reduction in venous return and hence cardiac output will also fall. Compensatory arteriolar constriction and eventually hypotension then lead to reduced organ and tissue perfusion (fluid flow through). This is hypovolaemic shock.

48

There are three vasomotor reflexes that can make cardiovascular adjustments within seconds. These are the baroreceptor, chemoreceptor, and ischaemic reflexes. All three can adapt, and are gradually desensitized over days if they are stimulated continuously. There are a further three mechanisms that can make cardiovascular adjustments over a period of minutes to hours. These are the transcapillary volume shift, vascular stress relaxation, and renin-angiotensin mechanisms. Finally, there are three systems that can be used to balance blood volume with vessel capacity in the longer term. These are the renal volume control, anti-diuretic hormone (ADH), and aldosterone systems.

Short-Term (Acute) Regulation:
Vessel capacity is adjusted to match the available blood volume by neural control. This adjustment is through the baroreceptor, chemoreceptor and ischaemic reflexes. There are numerous baroreceptors (stretch receptors or pressoreceptors) found in the walls of large arteries. The most significant are those of the aortic arch above the heart, and of the carotid sinuses in both the left and right carotid arteries at their bifurcations in the neck. These are depressant reflexes that decrease the heart rate (cf. the lovers' caress of the neck). Increases in either the amplitude or rate of change in pressure cause an increase in discharge frequency. Discharge leads to a depressor action; decreasing vasomotor tone and hence decreasing vasoconstriction, heart rate (HR) and the contractile force (determining stroke volume; SV) of the myocardium. Increasing discharge frequency leads to varying degrees of vasodilatation (vasodilation) in different vascular beds. Hence, there will be an increase in capacity of capacitance vessels and a decrease in total peripheral resistance (TPR); so arterial blood pressure (BP) will also decrease (BP = HR x SV x TPR). Overall then, an increase in blood pressure results in more stretch of the arteries, which due to

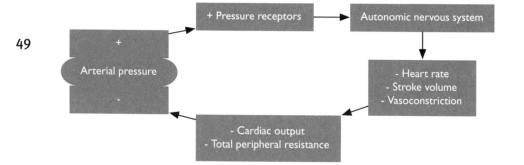

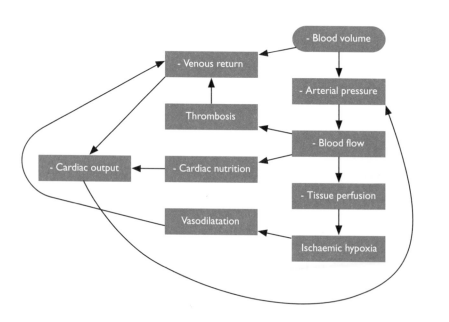

an increase in the continuous discharge rate triggers an increase in a number of depressor responses, leading to a decrease in arterial blood pressure. A decrease in blood pressure results in less stretch of the arteries, decreasing the continuous discharge rate, which triggers a decrease in a number of depressor responses, leading to an increase in arterial blood pressure. This negative feedback loop is the predominant reflex responsible for stabilising blood pressure. Baroreceptor activity also affects blood volume. Vasodilatation of arteries and arterioles conveys a greater pressure to the capillaries. An increased capillary pressure leads to an increased filtration from the capillary lumen into the interstitial space. Similarly, vasoconstriction leads to a decreased pressure and decreased filtration of intravascular fluid into the interstitial space. Thus the baroreceptor reflex can account for a change of 10-15 % in plasma volume over 15-20 minutes.

There are two types of stretch receptor that are found in both atria of the heart and nearby in the large veins. Type A receptors respond to atrial contraction, while type B receptors respond to passive atrial stretch during ventricular systole. Type A receptors respond to an increased blood volume which elicits an increased atrial pressure leading to an increased heart rate. This is particularly marked with rapid intravenous infusions and is called the Bainbridge effect (cf. Starling's law states that a higher venous return results in a stronger contraction.) Type B receptors respond to changes in pressure in a similar way to arterial receptors. Thus for example, a decreased pressure elicits less stretch and consequently an increased vasoconstriction in order to drive pressure back up. There is however one slight difference in the action of B receptors. In direct contrast to arterial receptors, the vasoconstriction action of B receptors is greater on renal vessels than muscle vessels. These B receptors are ideally placed to detect changes in intravascular volume via the dynamics of ventricular filling, and so they have a greater influence on the renal control of volume. B receptors also influence osmoregulation by stimulating the secretion of anti-diuretic hormone from the hypothalamus.

There are also numerous chemoreceptors (for oxygen, carbon dioxide and hydrogen ions) found in the walls of large arteries along with the baroreceptors. Systemic decreases in PO_2, increases in PCO_2, or increases in [H^+] will all increase stimulation of these chemoreceptors. Stimulation of carotid chemoreceptors leads to arteriolar

51

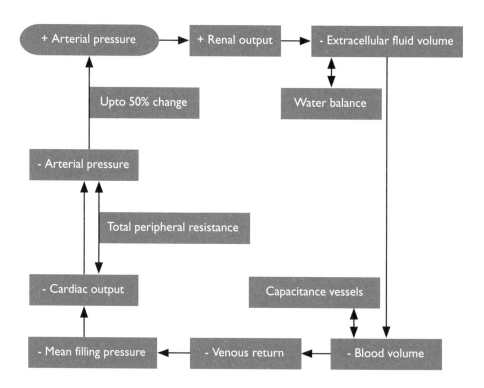

vasoconstriction and bradycardia. This causes a decrease in cardiac output, but blood pressure increases due to the proportionately larger increase in total peripheral resistance. Stimulation of aortic chemoreceptors leads to both arteriolar vasoconstriction and tachycardia, thus increasing both cardiac output and blood pressure. In both cases the increase in blood pressure then contributes to correcting the major systemic metabolic imbalances. However, these responses are greatly complicated by other responses which overlap at a local level. For instance, the diversion of blood flow by peripheral vasoconstriction also has a secondary affect on the kidneys. A reduction in renal perfusion results in a corresponding decrease in urine production, and increase in blood volume conservation, and hence further increasing blood pressure. In addition, regional decreases in PO_2 in the lung also cause vasoconstriction in those areas. Similarly, increases in PO_2 in the lung cause vasodilatation. This enables blood to be diverted to where most oxygen is available. However the opposite occurs in tissues elsewhere in the body, localized decreases in PO_2 cause vasodilatation in those tissues, and increases in PO_2 cause vasoconstriction. This enables correction of more minor localized metabolic imbalances.

Intermediate-Term Regulation:
Intravascular volume is interchanged with interstitial volume mostly by local control. This adjustment is through the transcapillary volume shift, vascular stress relaxation, and renin-angiotensin mechanisms. Transcapillary volume shift is the transfer of water from the intravascular space to the interstitial space, or vice versa. Any increase in capillary blood pressure leads to an increase in filtration. This causes an increase in interstitial volume and corresponding decrease in intravascular volume. As a direct consequence, the initiating pressure falls. Conversely, any decrease in capillary blood pressure leads to an increase in reabsorption and increase in intravascular volume.

Vascular stress relaxation refers to the delayed compliance of capacitance vessel walls in response to change in pressure. An increase in intravascular volume will increase pressure. After an initial stretching, the vessels continue to slowly distend over 10-60 minutes. The additional increase in intravascular volume causes a corresponding decrease in blood pressure. Conversely, with reverse stress relaxation, constriction to reduce intravascular volume causes an increase in blood pressure. This mechanism acts to return blood pressure to normal levels

53

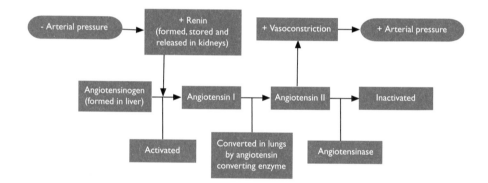

following sustained changes in intravascular volume.

Renin is an enzyme produced and stored in the kidney. A decrease in renal perfusion, for whatever reason, causes an increase in the secretion of renin. Renin is responsible for splitting angiotensinogen (which is produced in the liver) to give angiotensin I. Angiotensin I is converted to angiotensin II by angiotensin converting enzyme (ACE) in plasma. It is angiotensin II that causes an increase in total peripheral resistance (where angio = vessel, and tensin = tension; so angiotensin increases vessel tension). Angiotensin II increases vasoconstriction of arteries and to a lesser extent veins; it also stimulates the secretion of aldosterone from the adrenal cortex. Renin takes about 20 minutes to take effect through angiotensin II. Thus this mechanism applies if there is a sustained decrease in blood pressure or blood volume. Angiotensin II also stimulates thirst in an attempt to rectify low blood volume. (NB normal blood levels are: angiotensin II = 5-35 pmol.l^{-1}; aldosterone = 100-500 pmol.l^{-1}; and renin = 2.8-4.5 pmol.ml^{-1}.h^{-1} when erect, or 1.1-2.7 pmol.ml^{-1}.h^{-1} when supine and recumbent.)

Long-Term (Chronic) Regulation:
Renal excretion of water and electrolytes are adjusted by hormonal control to match blood volume with vessel capacity. This adjustment is through the renal volume control, anti-diuretic hormone, and aldosterone systems. The renal volume control system relies upon: changes in blood pressure affecting renal function, which in turn affects blood pressure. An increase in blood pressure leads to an increase in renal output. This causes a decrease in intravascular volume and a corresponding decrease in blood pressure (cf. the administration of diuretics to treat hypertension). Conversely, a decrease in blood pressure leads to a decrease in renal output, causing an increase in intravascular volume and a corresponding increase in blood pressure. Overall, there can be very large changes in renal output for apparently small changes in blood pressure. Hence, the first mmHg increase in blood pressure can double the volume of urine produced. A 10 mmHg rise can produce a six-fold increase in urine volume (polyuria); whereas a corresponding fall in blood pressure can severely decrease urine production (oliguria), or stop urine production altogether (anuria). The change in renal output can be transient or sustained according to requirements. This system is particularly important for maintaining an equilibrium between the intake and output of water by the body.

54

FOR WRITTEN NOTES:

The renal volume control system is further influenced by anti-diuretic hormone and aldosterone.

Anti-diuretic hormone (ADH or arginine vasopressin) causes vasoconstriction of arterioles and hence also a reduction in renal output. Normal blood levels are 0.9-4.6 pmol.l^{-1}. ADH is secreted by the pituitary gland in response to an increase in either osmolality or decrease in intravascular volume. ADH is therefore particularly involved in water regulation. An increase in intravascular volume leads to an increase in atrial receptor activity and consequently a decrease in ADH secretion, which causes an increase in renal output. Conversely, a decrease in intravascular volume leads to a decrease in atrial receptor activity and consequently an increase in ADH secretion, which causes a decrease in renal output. This system is particularly important for chronic changes in blood pressure, taking over the compensatory role from other baroreceptors.

Aldosterone causes an increase in the tubular reabsorption of sodium ions and water, and therefore via osmosis indirectly causes an increase in the output of potassium and hydrogen ions. Aldosterone is secreted by the adrenal cortex in response to a decrease in either sodium concentration (hyponatraemia) or intravascular volume. Aldosterone is therefore particularly involved in sodium regulation. Aldosterone increases the sodium and extracellular fluid content of the body. This occurs over a period of hours to days, and is closely linked with the renin-angiotensin system.

It is self evident that in order to regulate circulating blood volume there will be a rather complicated overlap of the nine responses outlined above.

HYPOVOLAEMIA AND HYPERVOLAEMIA:

Up to 500 ml of blood can be lost or gained, and within 15-30 minutes the change has been buffered by the interstitial spaces of the body. Larger changes can require much more time for compensation. Therefore, depending on the magnitude and rapidity, the body may not be able to compensate for changes in volume. However in many instances, volume imbalance can be artificially rectified if dealt with promptly.

For written notes:

57

Hypervolaemia (volume expansion) is an excessively high blood volume. Typically brought about by an increase of more than 500 ml. This increases blood pressure, causing overload of the circulatory system (i.e. preload), leading to heart failure, pulmonary oedema and acute respiratory distress syndrome (ARDS). It may be possible to remove the excess fluid volume by the administration of diuretics to increase urine production, or through the removal of blood by venepuncture.

Hypovolaemia (volume depletion) is an excessively low blood volume. Typically brought about by a decrease of more than 500 ml. This decreases blood pressure, causes inadequate perfusion, and leads to circulatory collapse and shock. It may be possible to replace lost blood volume by the transfusion of intravenous fluids.

SUMMARY OF VOLUME:

Volume defines the physical presence of a body. Volume regulation depends foremost on the control of temperature and water. Changes in body temperature directly affect body volume. An increase in temperature physically causes an expansion in volume. This is critical for the body, which not only has a hermetic limit, but also has a decreasing efficiency with increasing deviation from the optimum volume. Changes in body volume can also directly affect body temperature. For example, an increase in intracranial pressure from 10 mmHg to above 20 mmHg increases body temperature. Thermoregulation is therefore an essential component of volume regulation. Once a thermoregulated body plan has been established, determination of body volume on a day-to-day basis hinges on the control of pressure; particularly blood pressure. The cardiovascular system directly regulates the volume of fluid in the body, and in doing so it selectively influences the perfusion of other systems. The cardiovascular system particularly influences the renal system, but it also affects water loss through the skin and gastro-intestinal uptake. Thus although nutrition and excretion do require continuous fluctuations in body volume, it is the cardiovascular pressure that has an overriding influence. Aside from these effects, the role of the cardiovascular system in volume regulation also impinges on *all* other body processes. In turn, the cardiovascular pressure ultimately relies on the regulation of hydration.

For written notes:

59

CHAPTER 3.

HYDRATION

INTRODUCTION TO HYDRATION:

Hydration is the chemical combination of a substance, medium or body with the molecule of water (H_2O). Water is a prerequisite for life. The chemical and physical properties of water make it uniquely suitable for supporting all living processes. Liquid water is essential for the complex biochemical reactions that constitute life. Water is not only the medium of life, but in turn also provides the bulk of the volume for the structure of living things. The control of water depends on osmoregulation and thus electrolyte balance. With respect to life, osmoregulation is the control of fluid tonicity in a homoiosmotic body, ensuring osmostability and maintaining a constant internal pressure irrespective of external conditions. The difference in osmolality between the body and the environment is a delicate balance. Water is primarily gained macroscopically (i.e. by consumption), but there are additional gains from biochemical means (i.e. by metabolic oxidation). In aerobic respiration, the metabolism of each glucose molecule produces six water molecules via the electron transport system. On the macroscopic scale, water is lost primarily through the kidneys, gastro-intestinal tract, lungs and skin. Thus, the processes of nutrition, metabolism and excretion all play key roles in water regulation, and are supported primarily by the cardiovascular, gastro-intestinal and renal systems.

PHYSICS OF HYDRATION:

One important fact about water is that it is chemically *stable* at room temperature and pressure, and hence provides a dependable medium. The various roles of water in living things depend on its ability to *dissolve* substances (such as oxygen or nutrients) and its ability to *flow* (take blood for example). There are many physical properties to water that make it peculiarly suitable for living processes. In biological systems water is the universal solvent and provides a medium for transportation via aqueous solutions. Water has a low viscosity allowing rapid movement, while also having a comparatively high surface tension to provide cellular interfaces. A critical fact for life is that these roles are temperature dependent. Indeed, water has a high thermal conductivity

For written notes:

61

and so water can be used to readily transfer heat. However, water has a high specific heat capacity (e.g. requiring 4.2 kJ.kg^{-1} for a 1 degree rise when between 14.5 and 15.5 °C and at 101.325 kPa) and so is generally not susceptible to rapid changes in temperature. Because water loses and gains heat more slowly than any other liquid, water can transport heat from one part of the body to another via the blood (acting as a heat exchanger). As the body contains so much water, body temperature can be buffered against rapid changes. For this reason the vital chemical reactions going on inside the body are neither slowed nor accelerated to any great extent. There is one further anomalous property of water, which is that the maximum density of water occurs at 3.98 °C not at its freezing point. This affects both the order of solidification and relative volumes of the solid and liquid states. Thus, water freezes from the top down and ice floats, with important implications for aquatic life at these temperatures.

OSMOREGULATION:

Osmosis is the diffusion of a solvent (usually water) from a solution with a low solute concentration to a solution with a high solute concentration, through a differentially permeable membrane. There is a difference between osmolarity and osmolality. Osmolarity is a calculation with respect to the volume of solution and osmolality is a measurement with respect to the mass of the solvent. Osmolarity is the number of moles of solute in one litre of solution. This is a calculates the concentration of osmotically active particles irrespective of the substances that are involved, and is expressed in osmol.l^{-1}. In contrast, osmolality is the number of moles of solute in one kilogram of solvent. This involves the mass of the solvent and is the actual value measured, being derived from the mole of the substance, and is expressed in osmol.kg^{-1}. Where one mole (mol) of a substance contains 6.023×10^{23} elementary particles (Avogadro's constant), and is equivalent to the relative molecular mass in grams. Thus, osmolality takes into account the space occupied by the solute as well as the solvent. So, contrary to expectation osmol*ality* is the correct term to use (and not os*molar*ity) as most measurements are made by weight. Osmolality measurements enable the comparison of solute concentration, for example of blood and urine. Plasma osmolality is 275-305 mosmol.kg^{-1}, whereas urine osmolality is 50-1200 mosmol.kg^{-1}. Urine is normally hypertonic to body fluids due to reabsorption (resorption, where one material is

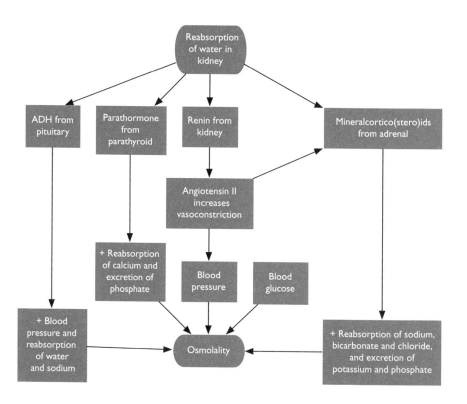

taken up by another) in the kidney tubules of much of the water that is filtered through the glomeruli. This is because of the overriding need for terrestrial life to conserve water. The elimination of water through the kidneys is primarily under hormonal control. There are four hormones directly affecting reabsorption from kidney tubules: ADH from the pars nervosa of the pituitary gland, parathormone from the parathyroid glands, renin from the juxta-glomerular complex cells of the kidney, and mineralocorticoids (mineralocorticosteroids, e.g. aldosterone) from the zona glomerulosa in the cortex of the adrenal gland. ADH increases reabsorption of water and sodium ions, and hence also increases blood pressure. Parathormone increases reabsorption of calcium and excretion of phosphate ions. Renin increases vasoconstriction by stimulating an increase in angiotensin II, and also indirectly increases the secretion of mineralocorticoids such as aldosterone. Together renin and mineralocorticoids increase reabsorption of sodium, bicarbonate and chloride ions; while increasing the loss of potassium and phosphate ions in the kidney tubules. In addition to these four hormones, factors (hormonal or otherwise) that affect blood glucose concentration will also affect osmolality. Similarly, factors that affect blood pressure will affect kidney filtration and hence affect osmolality. It follows that with water and ionic balance being constantly disturbed by the various losses and gains that occur with the processes of living, that osmoregulation is both highly complex and dynamic.

BODY WATER:

The volume of water and its distribution has serious implications for life. The proportion of the human body that is water declines with increasing age. For the young, 75 % of the body by volume is water, for adults it is 65 % (equivalent to about 42 l), and for the old-aged it is 55 %. Gender also affects water content. Males tend to have less fat and so are composed of proportionately more water. Any change in total body weight is usually accounted for by a change in extracellular water, fat (adipose tissue) or muscle. Total body weight is normally about 12 % extracellular water, 10 % fat, and 40 % muscle. The total proportion of body fat can vary from 10-50 %, and fat itself is composed of 10 % water. Total body fat and total body water are linked. Excluding fat, the mean water content of all body tissues is 73 %, hence:

% body fat = 100 - (% body water / 0.73)

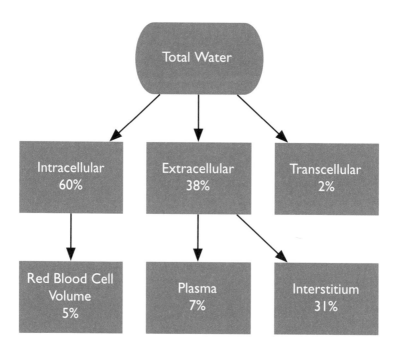

Thus if 10 % of an adult body is fat then 65 % of that body is water. Similarly, in the extreme case that fat accounts for 50 % of body weight, then the water content of that body is only 37 %.

BODY COMPARTMENTS:

The total water of the average adult is distributed between three body 'spaces' (compartments). These are the intracellular, extracellular and transcellular spaces. Cells contain most (60 % or 25 l) of this body water. For example, erythrocytes account for 2 l of total body water. That which is not intracellular (60 %) is either extracellular (38 %) or transcellular (2 %). About 13 l of extracellular water is in interstitial spaces, accounting for 31 % of total body water. About 3 l of extracellular water is in plasma, accounting for 7 % of total body water (cf. plasma is the acellular component of blood. Serum is plasma with all clotting factors such as fibrinogen removed, i.e. plasma clots and serum doesn't). Total intravascular water (or blood volume) is therefore 40 % intracellular and 60 % extracellular, accounting for about 12 % (5 l) of total body water. Transcellular water (the third space) accounts for 2 % (1 l) of total body water, and is found in cerebro-spinal fluid, the aqueous humour of eyes, gastro-intestinal lumen, glandular ducts, renal tubules and urinary ducts.

MOVEMENT ACROSS COMPARTMENTS:

It is possible to have redistribution of fluid from one body space to another. This is facilitated by changes in electrolyte balance. In general, cells are permeable to water but have low sodium permeability, and prevent the intracellular accumulation of sodium by active sodium-potassium ATPase transport outwards. Extracellular hypernatraemia will therefore draw water out of cells by osmosis. Thus, hyperosmolality leads to intracellular water moving to the extracellular space. On the other hand, hypo-osmolality leads to extracellular water moving to the intracellular space. Cell membranes are also relatively impermeable to intracellular proteins. The extracellular sodium and intracellular protein on opposite sides of the cell membrane remove the osmotic drive, preventing the movement of water. However, if the slow active outward transport of sodium is compromised (e.g. by hypoxia, cold or lack of substrate), then the cell swells. Increasing intracellular sodium also results in a corresponding decrease in intracellular potassium. A decline

Tissue Type	% Water	% Body Weight	Litres of Water in 70 kg Body
Blood	83.0	8.0	4.65
Kidneys	82.7	0.4	0.25
Heart	79.2	0.5	0.28
Lungs	79.0	0.7	0.39
Spleen	75.8	0.2	0.10
Muscle	75.6	41.7	22.10
Brain	74.8	2.0	1.05
Intestine	74.5	1.8	0.94
Skin	72.0	18.0	9.07
Skeleton	22.0	15.9	2.45
Adipose	10.0	10.8 (up to 50)	0.72 (up to 3.6)

in the potassium concentration gradient causes a reduction in the electrical charge across the membrane. This results in the membrane being increasingly permeable (e.g. to chloride ions). Eventually the membrane ceases to form an effective barrier and the cell dies.

Water can cross barriers by osmosis, or alternatively as part of an exudate or transudate. Exudation is the release of solution from the surface of the body. Exudation can occur physiologically such as with sweating (perspiration), or pathologically such as with burns where there is uncontrolled evaporation from the body surface similar to exaggerated sweating. On the other hand, transudation is the passage of a solution across a membrane within the body. Transudation can occur physiologically such as with capillary filtration, or pathologically such as with peritonitis where there is fluid leakage into the transcellular space of the abdomen.

WATER BALANCE:

Water intake and output must be balanced. Water output is through both sensible and insensible losses. There is normally an awareness of sensible loss and such losses can also be easily measured clinically, whereas there is not normally an awareness of insensible loss and such losses cannot be easily measured clinically. Thus sensible loss refers to urine (1500 ml.d^{-1}; approximately equivalent to 1 ml^{-1}.kg^{-1}.h^{-1}), whereas insensible losses comprise the water output through sweating and respiration (900 ml.d^{-1}; approximately equivalent to 0.5 ml^{-1}.kg^{-1}.h^{-1}) and faeces (100 ml.d^{-1}). Further unexpected losses are possible, such as through wounds (e.g. 3-10 ml.kg^{-1}.h^{-1} during surgery), vomiting (emesis) and diarrhoea. On the other hand, water is gained through drinking (1300 ml.d^{-1}), eating (900 ml.d^{-1}; bacon can be 10 % water and fruit 90 %, a mixed diet consists of about 60 % water) and oxidation (300 ml.d^{-1}; 0.6 ml.g^{-1} of carbohydrate oxidised, 1 ml.g^{-1} of fat, and 0.4 ml.g^{-1} of protein). Further artificial gains are possible, such as by intravenous infusion. The prevailing conditions can have a great effect on both sensible and insensible losses. Total sensible loss typically varies from 1-2 l.d^{-1}, whereas total insensible losses typically vary from 500-3500 ml.d^{-1}. In extreme cases, insensible water losses can reach as much as 20 l. However, water turnover is usually between 1.5 and 10 l.d^{-1}, which equates to about 4-24 % of total body water each day.

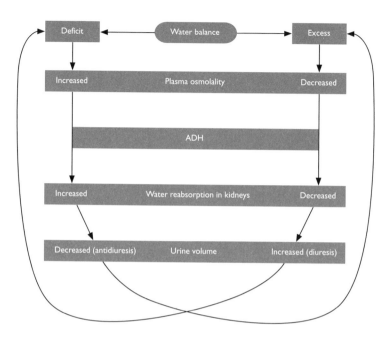

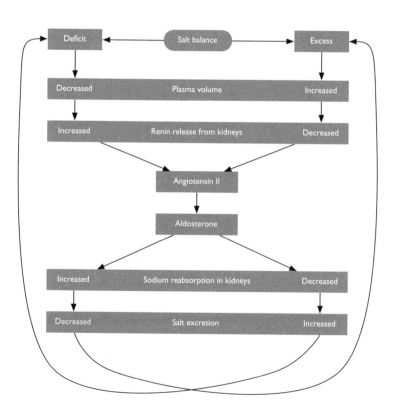

ROLE OF THE KIDNEY:

The structure and function of the kidney is extremely important for water balance. Each kidney comprises 6.17×10^5 nephrons. Each of these nephrons starts with a glomerulus, through which blood is filtered into a renal (Bowman's) capsule. Fluid passes from the capsule along a proximal tubule to a 'loop of Henle' and then a distal tubule before entering the collecting ducts. From here hypertonic urine passes from the ureter into the bladder for storage before elimination to the environment via the urethra. Secretion and reabsorption occur along the nephron to maintain osmolality and pH. Osmoreceptors (osmosensors) detect changes in body fluid osmolality (normally c. 290 mosmol.kg^{-1}). In turn the hypothalamus exercises hormonal control, by regulating the secretion of ADH, which affects renal elimination rate.

A water *im*balance may be due to either a deficit or excess, leading to *hyper*tonicity and *hypo*tonicity respectively. Hypertonicity can be caused by: sweating, hyperventilation, diuresis, chronic renal disease, or diabetes insipidus. In response to a 3 mosmol.kg^{-1} H_2O rise, cardiovascular pressure falls, and the concentration of angiotensin II rises in cerebro-spinal fluid. This leads to the secretion of ADH, and in turn thirst rises and renal excretion falls. A deficit of 4 % or more in total body water constitutes dehydration; this is associated with reduced salivation, dry mucous membranes, reduced skin turgor, and a rise in temperature. Whereas a loss of 4 % total body weight (due to dehydration) begins to impair performance. A loss of 20 % in total body water compromises cardiovascular function, whereas a loss of 20 % total body weight (due to dehydration) causes death. A potential consequence of dehydration is volume depletion (hypovolaemia), leading to circulatory shock and death. On the other hand, hypotonicity can be caused by: excessive drinking, intensive gastric lavage, or infusion of glucose solutions. In response, ADH secretion is inhibited and so more urine is produced. Thus in the long-term, excess water intake simply leads to elimination by the kidneys. However in the short-term, it is possible that the change can occur too quickly for ADH to be inhibited, leading to water intoxication. This can cause *intra*cellular oedema leading to cerebral swelling, and *extra*cellular oedema leading to restrictive pulmonary insufficiency. (NB oedema is swelling due to the accumulation of water.) In extreme cases nausea, vomiting, hypervolaemic shock and even death can occur.

70

FOR WRITTEN NOTES:

ROLE OF THE GASTRO-INTESTINAL TRACT:

The gastro-intestinal tract (GIT, gut or alimentary canal) also has an important role in water balance. The gastro-intestinal tract typically has an intake of 1.5 $l.d^{-1}$ via drinking. It also secretes about 7.5 $l.d^{-1}$; about 1500 ml of saliva, 2500 ml gastric juices, 1500ml pancreatic juices, 500 ml bile and 1500 ml of intestinal juices. Furthermore, the gastro-intestinal tract excretes approximately 100 $ml.d^{-1}$ through faeces in defecation. Overall, the gastro-intestinal tract therefore absorbs some 8.9 $l.d^{-1}$, nine-tenths of which is via the small intestine, with about one-tenth via the large intestine.

This absorption (cf. absorption is the process of penetration of one material by another, whereas adsorption is the action of one material being collected on the surface of another) is facilitated by electrolytes. Hydrochloric acid is secreted into the stomach, while sodium and bicarbonate ions enter the duodenum. This movement of electrolytes ensures that absorption of pure water does not occur predominantly by osmosis, but by isotonic reabsorption via sodium-potassium ATPase. This energy-dependent transport ensures a slow uptake, and thus prevents acute overloading of the cardiovascular system. The movement of electrolytes into the gastro-intestinal lumen initially causes the portal blood to become hypotonic. This blood is then transported to the liver, which responds by absorbing water (increasing the water content of the liver by up to 30 %). Osmoreceptors in the liver then signal the hypothalamus to decrease the secretion of ADH. The decrease in ADH causes an increase in renal output and a corresponding further decrease in intravascular volume. When the liver has become saturated, the osmolality of arterial blood begins to decrease. Small changes of 3 $mosmol.kg^{-1}$ can be detected by osmoreceptors in the hypothalamus itself. The hypothalamus has an overriding influence on ADH and hence reduces the diuresis. Intakes of isotonic solution do not stimulate osmoreceptors, but are directly regulated by cardiovascular responses to increased intravascular volume.

If there is a disturbance of the balance between the total amounts of secretion and absorption by the gastro-intestinal tract, the amount of water lost with faeces can vary dramatically. Hypercalcaemia causes constipation, whilst hypermagnesaemia causes diarrhoea. With diarrhoea liquid faeces are produced, usually because water is lost into

FOR WRITTEN NOTES:

73

the intestinal lumen osmotically. Many laxatives act via magnesium or a sulphate base, and cause water to enter the gut lumen and be lost at rates of up to 1 l.h^{-1} as diarrhoea (cf. cholera toxin acts by increasing cyclic adenosine monophosphate (cAMP) which in turn accelerates chloride ion secretion at the crypts of Lieberkuhn and again causes water to enter the gut lumen).

ROLE OF THE HYPOTHALAMUS:

Under normal conditions, the hypothalamus is more sensitive to changes in osmolality than volume. However, changes in blood volume exceeding 350 ml elicit an exponential increase in response. In extreme cases of both salt *and* water loss, the decrease in osmolality requires a decrease in ADH, while the decrease in volume requires an increase in ADH. Such a conflict is easily decided; the regulation of volume has priority. In the short-term it is more important to retain circulatory patency than to correct the dilution of blood. However, in the long-term volume cannot be maintained without adequate osmoregulation and electrolyte balance.

ELECTROLYTE BALANCE:

It is the overall integrity of the cell membrane that retains the cell as an independently living unit. The cell contains organelles dispersed in cytoplasm. Thus the cell contains particles, dispersed in an aqueous solution of water, ions and molecules. Cytoplasm is therefore colloidal. Colloid particles such as organelles are by definition immiscible, and typically have a molecular weight greater than 30,000. The particles are therefore suspended indefinitely and their surface area provides an important anchorage for a variety of reactions. The concentration of such particles can have an effect on osmosis. Thus the particles produce a colloid osmotic (oncotic) pressure, that along with the crystalloid osmotic pressure (due to electrolytes), combines to give the total osmotic pressure. Most of the oncotic pressure within cells is due to organelles, while the oncotic pressure outside cells is due to proteins such as albumin. Differences in protein distribution therefore affect oncotic pressure and hence water movement across body compartments. For example, interstitial oncotic pressure is 0.7 kPa, whereas plasma oncotic pressure is 3.3 kPa, giving a resting oncotic gradient of 2.6 kPa. Therefore, a decrease in plasma albumin will

For written notes:

75

reduce the oncotic pressure of plasma and lead to interstitial oedema. The particles also tend to be between 1 nm and 1 μm in diameter (e.g. albumin is 8 x 3 nm) and so cannot pass through cell membranes. Colloid particles typically account for only 0.5 % of total osmotic pressure, but the fact that they can not cross capillary membranes means that they are an important factor in transcapillary fluid dynamics. Water is continuous with both the inside and outside of cell membranes; and although colloid particles cannot cross cell membranes, solutes such as electrolytes can. Therefore, it is necessary for an active exchange of solutes across the membrane in order to maintain an optimum amount of water within the cell. The regulation of electrolyte transfer across cell membranes is therefore particularly important for the movement of water. This is necessary for the cell to survive and function efficiently.

Most cells are bathed in a solution with similar osmotic properties to weak sea water. Deviations result in either dilution or concentration of body fluids, both of which can result in serious problems for metabolism. Consider erythrocytes that are normally biconcave and discoid in shape. Erythrocytes placed in a hypertonic solution decrease in volume by extrusion of water and become crenated (shrivelled). Whereas, erythrocytes placed in a hypotonic solution increase in volume by the osmotic absorption of water and become haemolysed (ruptured). These changes are due to a net flow of water across the corpuscle membrane. Either case has obvious deleterious consequences for the efficient transport of oxygen by the circulation. For there to be no net flow of water and so no change in volume, an isotonic solution (in this case equivalent to about 1.7 % sodium chloride; 17 g.l^{-1}) is needed.

Ionic Moieties:
For neutrality, the electrical charge of all ions must be balanced by other ions that have an equal but opposite electrical charge. Each ion is therefore one half of the whole, i.e. it is a moiety. Certain ionic moieties are of greater physiological significance than others and so they are considered in isolation. In particular, the maintenance of volume depends on fluid and electrolyte balance. The electrolytes of significance here are: sodium (Na^+), potassium (K^+), calcium (Ca^{2+}), magnesium (Mg^{2+}) and phosphate (PO_4^{3-}). A deficit or excess of any of these has consequences for volume, primarily because of their effect on fluid movement. Deficits are given the prefix 'hypo' and excesses are

	NEEDS MMOL.D^{-1}	CONCENTRATIONS MMOL.L^{-1}			
	Total	Intra-	Extra-	Hypo-	Hyper-
Na$^+$ -natr-	50-100	10	142	<130	>150
K$^+$ -kal-	70-140	155	4	<3	>5
Ca^{2+} -calc-	25	0.001	2.5	<2	>3
Mg^{2+} -magnes-	15	15	0.9	<0.6	>1.6
PO$_4$$^{3-}$ -phosphat-	35	65	1	<0.6	>1.7

given the prefix 'hyper', while an ion imbalance in the blood is given the suffix 'aemia'.

Sodium is predominantly an extracellular cation. There is about 4.2 moles of sodium within the whole body. Approximately 10 mmol.l^{-1} (2.5-6 %) is found intracellularly, 142 mmol.l^{-1} (50-60 %) extracellularly, and most of the remainder (about one-third) is bound in bone. The remainder is in equilibrium with plasma, thus any losses from the extracellular space are partially compensated for by diffusion. The daily requirement for sodium is about 50-100 mmol. Twice as much is consumed by the average Western European, and is associated with high blood pressure. Hypernatraemia corresponds to an excess of Na$^+$ in serum >150 mmol.l^{-1}. Whereas, hyponatraemia corresponds to a deficit of Na$^+$ in serum <130 mmol.l^{-1}.

Potassium is predominantly an intracellular cation, important for cardiovascular and neuromuscular function. There is about 3.3 moles of potassium within the whole body. Approximately 4 mmol.l^{-1} is found extracellularly, and 155 mmol.l^{-1} intracellularly. Thus only about 3 % is outside cells, but potassium found inside cells is freely exchangeable if this balance is disrupted. Should there be a fall in extracellular potassium, then diffusion from the intracellular space compensates. Therefore the level of potassium in plasma is relatively stable and does not reflect intracellular levels. For example, a loss of 10 % of body potassium only causes the level of plasma potassium to fall by 1 mmol.l^{-1}. The daily requirement for potassium is about 70-140 mmol. Hyperkalaemia corresponds to an excess of K$^+$ in serum >5 mmol.l^{-1}. Whereas, hypokalaemia corresponds to a deficit of K$^+$ in serum <3 mmol.l^{-1}.

Calcium the most abundant cation in the body, and is mostly found as a component of bones and teeth. There is about 28 moles of calcium within the whole body. Approximately 2.5 mmol.l^{-1} (1 %) is found extracellularly, and 0.001 mmol.l^{-1} intracellularly, thus there is a steep concentration gradient across cell membranes. Calcium is actively transported out of cells, opposing a steady influx due to the large extracellular calcium concentration. Changes in intracellular calcium concentration are a crucial signal for many functions (e.g. excitation-contraction coupling in muscles). Approximately 49 % of extracellular calcium is bound to protein (mostly albumin), 1 % is bound to organic

For written notes:

79

acids and 49 % is ionised. Acidosis will increase the amount of ionised calcium, whereas alkalosis will decrease the amount of ionised calcium. The daily intake of calcium needs to be about 25 mmol, as only about 25 % is absorbed. Hypercalcaemia corresponds to an excess of Ca^{2+} in serum >3 mmol.l^{-1} (>3.5 mmol.l^{-1} can cause cardiac arrest). Whereas, hypocalcaemia corresponds to a deficit of Ca^{2+} in serum <2 mmol.l^{-1}.

There is about 1 mole of magnesium within the whole body. Bone accounts for 53 %, while muscle and other soft tissues account for 46 %. Of the remaining magnesium, approximately 15 mmol.l^{-1} is found intracellularly (in erythrocytes), and 0.9 mmol.l^{-1} extracellularly. Magnesium tends to move in and out of cells with calcium or potassium. The daily requirement for magnesium is about 15 mmol. Hypermagnesaemia corresponds to an excess of Mg^{2+} in serum >1.6 mmol.l^{-1}. Whereas, hypomagnesaemia corresponds to a deficit of Mg^{2+} in serum <0.6 mmol.l^{-1}.

There is about 23 moles of phosphorus within the whole body, bound as phosphate. Approximately 1 mmol.l^{-1} is found extracellularly, and 65 mmol.l^{-1} intracellularly. Most of the phosphate in the body (approximately 80 %) is found as a component of bone (partly associated with calcium), but it has many other important uses such as in ATP, DNA, membranes (in phospholipids), and reacting with hydrogen ions for their elimination (for pH regulation). The daily requirement for phosphate is about 35 mmol. Hyperphosphataemia corresponds to an excess of PO_4^{3-} in serum >1.7 mmol.l^{-1}. Whereas, hypophosphataemia corresponds to a deficit of PO_4^{3-} in serum <0.6 mmol.l^{-1}.

Other important components of serum include: chloride (95-110 mmol.l^{-1}; the principal anion), bicarbonate (24-30 mmol.l^{-1}; a principal pH buffer), cholesterol (3.9-7.8 mmol.l^{-1}; a product of lipid metabolism and a constituent of all steroid hormones), urea (2.5-6.7 mmol.l^{-1}; an end product of protein metabolism), glucose (3.3-5.5 mmol.l^{-1}; a source of energy), creatinine (40-150 μmol.l^{-1}; an end product of muscle metabolism), bilirubin (3-25 μmol.l^{-1}; an end product of haemoglobin breakdown and incidentally responsible for the brown colour of faeces), cortisol (80-700 nmol.l^{-1}; a hormone which increases blood glucose while decreasing inflammation and repair), and albumin (35-50 g.l^{-1}; a protein contributing 80 % of colloid osmotic pressure, and a large surface area for adsorption e.g. of bilirubin).

FOR WRITTEN NOTES:

81

Dehydration and Hyperhydration:

A water deficit is referred to as dehydration, whereas an excess of water is referred to as hyperhydration. Exchanges of water occur across the extracellular space. All such exchanges of water occur with reference to a normal plasma osmolality of 290 mosmol.kg^{-1}. Isotonic exchange occurs at this concentration, hypotonic exchange occurs at lower extracellular concentrations, and hypertonic exchange occurs at higher extracellular concentrations. There are thus six possible ways to disrupt the equilibrium of water.

Isotonic dehydration occurs due to the loss of both water and electrolytes equivalent in osmolality to extracellular fluid. This is usually caused by the loss of either whole extracellular fluid or whole transcellular fluid. Extracellular losses can be due to bleeding (haemorrhage), seepage following burns, or blood donation. Transcellular losses can be due to excessive sweating, vomiting, or diarrhoea. In either case, there is no change in the remaining extracellular concentration and so no osmotic effect on intracellular concentration; thus cell volume does not change. There is however a decrease in blood volume that also causes a decrease in blood pressure. Extreme cases lead to syncopy (fainting), hypovolaemic shock, and circulatory collapse.

Hypotonic dehydration occurs due to the output of electrolytes without sufficient water. The retention of water in this way can be caused by cardiac insufficiency or acute renal failure. The associated decrease in extracellular concentration leads to an increase in intracellular volume. Extreme cases lead to cerebral oedema.

Hypertonic dehydration can occur due to loss of too much water compared with the loss of electrolytes. This can be caused by an inadequate activation of thirst, heat adapted sweating, or diabetes. In such cases there is a decrease in both extracellular volume and intracellular volume. Inadequate thirst can occur at high altitude due to oxygen insufficiency, or when hyperventilating in the cold due to large unrecognised water losses in exhaled air. Heat adapted sweating produces comparatively hyponatraemic perspiration caused by the increased action of aldosterone on the sudoriparous (sweat) glands. Diabetes insipidus produces hypotonic urine caused by the decreased

FOR WRITTEN NOTES:

83

action of ADH. Diabetes mellitus produces isotonic urine, but because this is mostly attributed to glucose there is effectively retention of electrolytes and an osmotic diuresis. Extreme cases are associated with worsening extracellular and intracellular hypovolaemia.

Isotonic hyperhydration occurs due to the gain of both water and electrolytes equivalent in osmolality to extracellular fluid. This can be caused by renal insufficiency or hyperaldosteronism. As there is no change in the concentration of the expanding extracellular volume, there is no osmotic effect on intracellular concentration and cell volume does not change. There is however an increase in blood volume that also causes an increase in blood pressure. This is how inadequate sodium excretion can lead to oedema. The retention of sodium also causes the retention of water in order to maintain extracellular isotonicity. Extreme cases are associated with worsening hypervolaemia and hypertension.

Hypotonic hyperhydration (water intoxication) occurs due to the intake of water without sufficient electrolytes. This can be caused by drinking pure water in response to isotonic dehydration. The result is the same as for hypotonic dehydration. A decrease in extracellular concentration leads to an increase in intracellular volume. Extreme cases lead to cerebral oedema.

Hypertonic hyperhydration occurs due to the intake of water with excess electrolytes. This can be caused by drinking sea water (10^3 mosmol.kg^{-1}, salinity = 35×10^3 ppm or 35×10^3 mg.l^{-1}, density = 1.025 kg.m^{-3}) or intravenous infusion. As with hypertonic dehydration, an increase in extracellular concentration leads to a decrease in intracellular volume. For the kidneys to excrete the excess electrolytes, additional water from the body must also be excreted; leading to hypertonic dehydration.

SUMMARY OF HYDRATION:

Water provides the pressure to maintain boundaries and composes the bulk of body volume. Water permits the dissolution of gases and metabolites, providing a transport medium. Water also provides a reactive medium for manipulating energy, and even providing a source of pH buffering (as H_2O gives H^+ and OH^-). The regulation of hydration is therefore critical, providing the media within which all other body processes will occur.

For written notes:

85

CHAPTER 4.

OXYGENATION

Introduction to Oxygenation:

Oxygen (O_2) is the most abundant and widespread of all the elements in nature. Approximately, one-fifth (20.93 %) of the volume of the air around us is pure oxygen, nine-tenths of the weight of water consists of chemically bound oxygen, and one-half of the weight of the Earth's crust is oxygen bound up as solid oxides. Oxygenation is the chemical combination of a substance, medium or body with the element of oxygen. Gaseous exchange is critical for life. For animal life, oxygen needs to be extracted from (and carbon dioxide exchanged with) a fluid environment in order for energy to be derived from a variety of substrates. The key processes are: the movements which facilitate gaseous exchange, and the transportation of gases within the liquid of body fluids. Hence, it is the respiratory and cardiovascular systems that are involved.

Physics of Oxygenation:

The principal functions of the respiratory system are that of transport and exchange. The respiratory system serves to ensure that all tissues receive the oxygen they need, and can dispose of the carbon dioxide they produce. Blood carries gases to and from the tissues, while the lungs exchange gases with atmosphere. Blood has the intrinsic capacity to pick up oxygen and lose carbon dioxide. However, such exchanges can only occur if blood is exposed to the right gaseous environment, and this is achieved by the lungs.

From the kinetic theory of gases, gases are thought of as a collection of molecules moving around a space. Pressure is generated by collisions of molecules with the walls of that space. The more frequent and harder the collisions the higher the pressure. From Boyle's law, if a given amount of gas is compressed into a smaller volume, molecules will hit the wall more often and pressure will rise. So pressure is inversely proportional to volume. From Charles' law, the kinetic energy of molecules increases with temperature. As temperature increases, molecules hit the walls more often and pressure increases. So pressure

For written notes:

87

is proportional to absolute temperature. The Universal (ideal) gas law allows the calculation of how volume will change as pressure and temperature changes:

$$P.V = n.R.T.$$

Where, pressure is in kPa, volume in l, n = number of moles of gas, R (molar gas constant) = 8.3145 J.K^{-1}.mol^{-1}, and temperature in Kelvin. Volumes are usually corrected to standard temperature and pressure (STP; equal to 273.15 K and 101.325 kPa).

In a mixture of gases molecules of each type behave independently, so each gas exerts a *partial pressure*. From Dalton's law, the partial pressure of individual gases adds up to the total gas pressure. Partial pressures are calculated as the same fraction of the total pressure as the volume fraction of the gas in the mixture.

For example, in air the partial pressure of oxygen (PO_2):
= total pressure x 0.2093
= 101.325 X 0.2093
= 21.177 kPa

In biological systems gas mixtures are always in contact with water. So water molecules evaporate and gas molecules dissolve. Water molecules entering the gas exert a vapour pressure. Saturated vapour pressure is when molecules leave and enter at the same rate, and depends only on temperature (6.28 kPa at 37 °C).

Exchange between air and blood occurs across the alveolar membrane. The surface area of alveoli potentially available for gaseous exchange is about 100 m^2 (try to visualize half a tennis court fitting inside the thorax). 'Alveolar air' has a different gaseous composition to the atmosphere. There is less oxygen and more carbon dioxide. For alveolar air: P_AO_2 is normally 13.3 kPa, and P_ACO_2 is normally 5.3 kPa. Mixed venous blood returns to the lungs from the body, where P_vO_2 is typically 6.0 kPa and P_vCO_2 is typically 6.5 kPa, although these values vary with metabolism. When considering the gradients of partial pressure: P_AO_2 in alveolar gas > P_vO_2 in returning blood, and P_ACO_2 in alveolar gas < P_vCO_2 in returning blood. The exchange occurs by diffusion, with gases moving along their concentration gradients. So oxygen diffuses

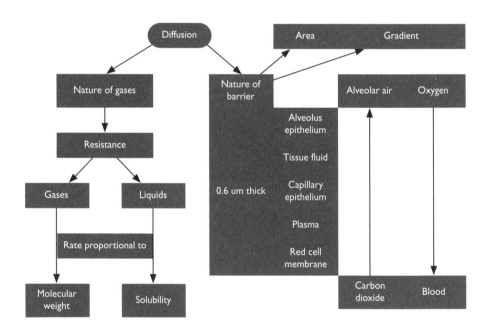

into the blood and carbon dioxide diffuses out of the blood. Increased diffusion requires a larger area, larger gradient, and / or smaller resistance. Diffusion resistance depends on the nature of barrier and the nature of the gas(es). In the lungs, the diffusion barrier consists of the: epithelial cell of the alveolus, tissue fluid, endothelial cell of the capillary, plasma, and red cell membrane. The diffusion barrier therefore comprises: five cell membranes, three layers of cytoplasm and two layers of tissue fluid. Diffusion of gases through gases is at a rate inversely proportional to molecular weight. Larger molecules diffuse more slowly. So carbon dioxide is slower than oxygen. However, diffusion of gases through liquids is at a rate proportional to solubility. Carbon dioxide is much more soluble than oxygen. Thus despite being a larger molecule, carbon dioxide diffuses 21 times faster than oxygen. So exchange of oxygen is always limiting. Overall, the diffusion resistance barrier is 0.6 μm thick (or rather 0.6 μm *thin*). An erythrocyte has almost one second of contact time within an alveolar capillary. This is twice as long as required for complete oxygen exchange by diffusion. So, gas diffusion does *not* limit respiratory function.

CHEMISTRY OF OXYGEN IN THE BODY:

Tissue respiration not only includes the transport and exchange of gases, but also the oxidation of nutrients. This ultimately facilitates the storage of energy in the chemical bonds of ATP. In aerobic respiration, the metabolism of each glucose molecule produces six carbon dioxide molecules. Of the 38 ATP molecules produced from the metabolism of each glucose molecule, only two ATP molecules (about 5 %) are synthesized without oxygen. Thus oxidative phosphorylation yields about 95 % of ATP molecules, whilst the remainder is derived via substrate-level phosphorylation. The breakdown of glucose under aerobic conditions yields about 2826 kJ.mol^{-1}, whereas the breakdown of glucose under anaerobic conditions (glycolysis) yields only about 208 kJ.mol^{-1} (along with an accumulation of lactate).

Oxygen is also a free radical. A free radical is a molecule with one or more unpaired electrons in its outer orbital. Such molecules are relatively unstable and are much more likely to react with the outer orbital electrons of other molecules. If the unpaired electron is passed on to another molecule, then there can also be a chain reaction of free radical generation, as electrons are lost and gained from one molecule

For written notes:

to the next. The metabolism of toxic chemicals can also generate
free radicals. Oxygen free radicals (reactive oxygen species) can cause
several types of cellular damage (oxidative stress). Reactive oxygen
species are implicated in: carcinogenesis (if the damage is to regions
of DNA regulating cell proliferation), radiation injury (two-thirds of such
damage is due to the secondary generation of reactive oxygen species),
atherosclerosis (due to peroxidation of low density lipoproteins;
forming fat plaques on artery walls), reperfusion injury (due to increased
oxygen following ischaemia; an important factor for organ transplants
and multiple organ failure), and ageing (where cell membrane
poly-unsaturated fatty acids are progressively oxidised). Such oxidative
stress can be offset by some molecules acting as antioxidants
(e.g. vitamin E, glutathione, or the enzyme superoxide dismutase),
scavenging free radicals by reacting preferentially with reactive oxygen
species without passing that reactivity along. However, the reactivity
of some reactive oxygen species has been harnessed by various
cellular processes (e.g. lysosomes recycling cellular components, and
macrophages engulfing and neutralizing foreign material). Thus the
presence of reactive oxygen species is also important for the
maintenance of normal function.

CLINICAL MEASUREMENT OF OXYGEN:

The analysis of blood gases most commonly depends on the invasive
approach of taking a blood sample (and is covered under the 'Clinical
Measurement of pH'). Non-invasive methods are more technologically
sophisticated and hence expensive by comparison. However
spectroscopic probes such as transdermal pulse oximeters allow
real-time monitoring, and so can be especially useful in certain trauma,
oncology and cardiovascular cases. As oxygen is transported in blood
mainly by haemoglobin, the total amount of haemoglobin and what
proportion of that haemoglobin has bound oxygen are both important.
Thus oxygen saturation is frequently measured along with the
concentrations of total oxygenated and deoxygenated haemoglobin, and
considered in conjunction with cardiac output and fluid volume status.

NORMAL BODY OXYGEN:

The oxygen uptake by the lungs directly replaces that removed from
the blood by the tissues. The oxygen uptake of the body is

FOR WRITTEN NOTES:

200-250 ml.min^{-1} at rest, increasing to 600-700 ml.min^{-1} during moderate activity, and to about 1 l.min^{-1} during extreme exercise. However, oxygen requirement varies with tissue type: skeletal muscle requires 2.5-5.0 x 10^{-3} ml.g^{-1}.min^{-1} (0.1-0.2 ml.g^{-1}.min^{-1} during extreme exercise), heart requires 7-10 x 10^{-2} ml.g^{-1}.min^{-1} (40 x 10^{-2} ml.g^{-1}.min^{-1} during extreme exercise), brain requires 3-4 x 10^{-2} ml.g^{-1}.min^{-1}, liver requires 5-6 x 10^{-2} ml.g^{-1}.min^{-1}, and kidney requires 5.5-6.5 x 10^{-2} ml.g^{-1}.min^{-1}, while blood itself consumes 0.6-1.0 x 10^{-4} ml.g^{-1}.min^{-1}.

One mole of haemoglobin (Hb) is capable of binding a maximum of 4 moles of oxygen: Hb + 4O$_2$ reacts to give Hb(O$_2$)$_4$

The molecular weight of haemoglobin is 64,500, and thus 64.5 kg of haemoglobin could potentially bind 4 x 22.4 l of oxygen (so 1 g of haemoglobin can bind 1.39 ml of oxygen). Practically, not all of haemoglobin in blood is available for binding. So 1 g of haemoglobin actually binds 1.34-1.36 ml of oxygen. There is about 150 g of haemoglobin in one litre of blood, and so it follows that the maximum oxygen carrying capacity of blood is about 200 ml.l^{-1}. An oxygen uptake of 200 ml.min^{-1} would therefore only require a pulmonary perfusion of 1 l.min^{-1}. Nevertheless, the total cardiac output passes through the lungs. A total cardiac output of 5 l.min^{-1} therefore permits a maximum oxygen uptake of one litre. Thus at rest the oxygen requirement of the whole body is only about 20 % of total capacity.

Thus overall the body consumes about 20 % of the available oxygen in blood at rest. However, any degree of ischaemia will cause a proportional increase in utilization. Utilization of oxygen also varies with different types of tissue. For example it is much lower in the kidney, because of the comparatively large volumes of blood flow through the tissue. Utilization can also be much higher, as the body can consume up to 90 % of available oxygen during extreme exercise. During exercise, oxygen consumption by skeletal muscle can increase by about 30-40 fold, and cardiac muscle by 3-4 fold above the resting level. Within muscle cells, myoglobin acts to transport and store oxygen. By reversibly binding oxygen, myoglobin acts as a short-term oxygen buffer. This enables continuous work when the oxygen supply is interrupted. Following such an interruption, oxidative metabolism can continue for 3-4 seconds. This is important for the myocardium when

95

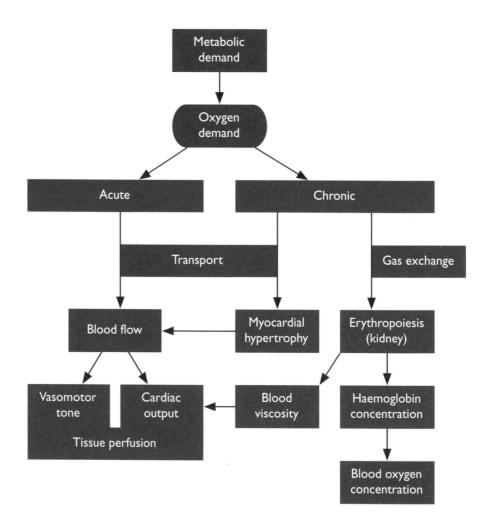

oxygen availability is low during systole, enabling the maintenance of efficient heart rates of up to 200 beats.min^{-1}. Myoglobin is similarly important for skeletal muscle, maintaining the efficiency of contraction during the transition to an increased workload.

CHANGES IN OXYGEN DEMAND:

An increase in metabolic rate demands a corresponding increase in the requirement for oxygen. An increase in oxygen demand, which exceeds the reserves of oxygen normally available, can only be achieved by increasing either exchange or transport. However, there is not much scope for acute increases in exchange. The saturation of haemoglobin with oxygen is normally about 97 %, so increasing the oxygen concentration in arterial blood is impractical. Therefore, any acute increase in oxygen requirement must depend on increasing transport, i.e. increasing blood flow. Increases in blood flow can be brought about by changes in either cardiac output or vasomotor tone. Thus an increase in oxygen demand is met by an increase in the perfusion of the tissue above the level required for essential maintenance. Once exposure to an adequate PO_2 is established, further increases in blood flow are ineffective. This is because the rate of oxygen uptake from the blood is independent of the rate of blood flow through the tissue. In other words, the constant oxygen gradient provided by a continuous flow of blood past a stationary tissue is sufficient for exchange to occur. (NB the kidney is an exception to this, as an increase in blood flow will cause an increase in filtration rate and hence also an increase in oxygen demand.)

Sustained increases in demand result in growth of the heart muscle cells (myocardial hypertrophy), whereby the adult heart can double in weight to reach 500 g. This growth is limited by the fixed capacity of the pre-capillary coronary vessels that supply the blood to the heart. In addition to myocardial hypertrophy, there can also be an increase in erythropoiesis and hence increase in haemoglobin concentration. Although this does not change PO_2, it does increase the oxygen concentration in blood. Thus the oxygen carrying capacity of blood is increased. However, with increasing numbers of erythrocytes the blood viscosity also increases. This increasing viscosity requires an increasing cardiac output in order to maintain blood flow. Thus the capacity to increase blood flow is limited.

97

CLASSIFICATION OF RESPIRATORY DISEASE:

Respiratory diseases are classified as restrictive or obstructive (or a combination of both). People are also sometimes described by their appearance, as either a 'blue bloater' or a 'pink puffer'.

Someone with a restrictive respiratory deficit has restricted lung expansion due to decreased compliance (e.g. caused by fibrosis). So they have difficulty breathing in, but they have no difficulty breathing out. Thus there is a decrease in lung volume and often a decrease in gas exchange. This can then lead to type II respiratory failure (ventilatory failure) which is defined as the presence of both hypoxaemia and hypercapnia (i.e. there is carbon dioxide retention or trapping); as opposed to type I respiratory failure which is defined by the presence of hypoxaemia alone without hypercapnia (e.g. due to high altitude when there is no reduction in carbon dioxide exhaled). Decreased effectiveness of ventilation leads to decreased exercise tolerance and weight gain. Also, sustained attempts to increase lung perfusion via pulmonary hypertension leads to enlargement of the right ventricle and right-sided heart failure (cor pulmonale), which then leads to peripheral oedema. Furthermore, ventilatory failure leads to chronic hypoxaemia and polycythaemia. The increased proportion of deoxygenated blood relative to oxygenated blood leads to cyanosis (a cyan or bluish complexion). Cyanosis, oedema and weight gain gives rise to being described as a 'blue bloater'.

Someone with an obstructive respiratory deficit has obstructed air flow due to increased airway resistance (e.g. caused by emphysema). So they have no difficulty breathing in, but they do have difficulty breathing out. This is because at rest inhalation is active and lung expansion is not impaired, whereas exhalation is passive (occurring due to elastic recoil when inspiratory muscles relax) and so depends more on airway resistance. This can lead to persistent and progressive breathlessness, causing rapid and short breaths from overinflated lungs (barrel chest). Sustained hyperventilation leading to a reddish complexion and weight loss gives rise to being described as a 'pink puffer'.

Local increases in metabolic rate also lead to increases in the dissociation of oxygen from haemoglobin. Where increased metabolism produces more carbon dioxide, pH decreases. As pH decreases, the affinity of haemoglobin to bind with oxygen decreases, so more oxygen is released. This is called the Bohr effect. The opposite occurs in the lungs when carbon dioxide is breathed out and pH increases.

Changes in body temperature will also change metabolic rate and hence also change oxygen demand. In theory, a decrease from 37 °C to 27 °C would cause about a 2-3 fold decrease in oxygen demand. However under normal circumstances, thermoregulatory responses will actually maintain body temperature, by increasing metabolic rate and hence actually increasing oxygen demand.

HYPOXIA AND HYPEROXIA:

Hypoxia is an excessively low PO_2 (cf. hypoxaemia is an excessively low blood oxygen concentration). This is in contrast to anoxia, which is the total absence of oxygen. Hypoxia occurs if PO_2 in alveoli falls below 4.7 kPa. There are four principal types: hypoxic, anaemic, ischaemic and histotoxic. Hypoxic hypoxia (arterial or pure hypoxia) is due to a decrease in the oxygen reaching the red blood cells. Hypoxic hypoxia can be caused by respiratory disease (e.g. emphysema, pulmonary oedema, pneumonia, fibrosis), trauma (e.g. pneumothorax causing the lung to collapse), obstruction (which can be due to disease or trauma), hypoventilation, or a failure of acclimatization to elevated altitude. Anaemic hypoxia is due to a decrease in the oxygen carrying capacity of blood, e.g. caused by sickle cell disease. Ischaemic hypoxia is due to inadequate perfusion, e.g. caused by peripheral vascular disease. Histotoxic hypoxia is due to competition with oxygen for binding to haemoglobin, e.g. caused by cyanide or carbon monoxide poisoning.

Hypoxic hypoxia is a deficit in PO_2, and can occur without a change in PCO_2. This occurs with some diseases and also on exposure to high altitude. With increasing altitude, total pressure falls. The volume fraction of oxygen remains the same (20.93 %), so PO_2 falls. At sea level, PO_2 in air is 21.2 kPa, while in alveoli it is 13.3 kPa. The critical level of alveolar P_AO_2 needed to maintain respiration is 4.7 kPa, which is equivalent to an altitude of 4 km (where PO_2 in air = 10 kPa). The symptoms of pure hypoxia are like those of alcohol. Initially there

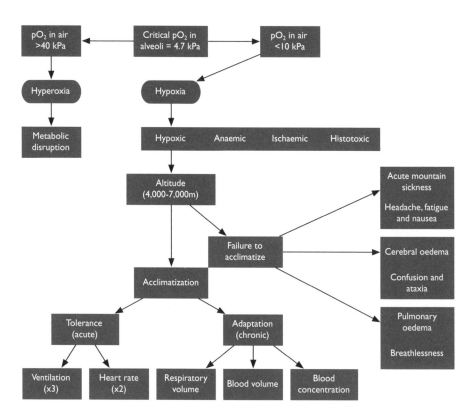

is a pleasant euphoria, then nausea, inco-ordination, unconsciousness and death. Indeed, hypoxiphilics use asphyxia (producing hypoxia and hypercapnia; i.e. low PO_2 and high PCO_2) for sexual gratification, and excesses have caused some accidental deaths. (NB hypercapnia without hypoxia also leads to unconsciousness. Deterioration in performance begins when inhaled levels of carbon dioxide reach 3-5 % and unconsciousness occurs when levels exceed 15 %.)

Acclimatization to changes in environmental oxygen can take the form of tolerance or adaptation. Tolerance involves short-term tactics, while adaptation involves long-term strategies. With acute exposure to very high altitude (above 4 km), ventilation and heart rate are stimulated to increase. This is limited to a three-fold increase in respiration rate and two-fold increase in heart rate. These tactics maintain alveolar P_AO_2 above 4.7 kPa up to 7 km (where PO_2 in air = 6 kPa). Above 7 km continued exposure will cause deterioration in awareness and eventually lead to unconsciousness. Acute exposure can occur due to the depressurization of an aeroplane at high altitude. With acute exposure to 7 km useful consciousness lasts for about five minutes, whereas at 14 km useful consciousness lasts for only 10-20 seconds. Breathing 100 % oxygen permits normal ventilation up to 12 km (where PO_2 in air = 2 kPa). Coupling 100 % oxygen with increasing ventilation, it is possible to extend the range for survival up to a maximum of 14 km (where PO_2 in air = 1 kPa). Without pressurization, increasing altitude eventually results in a physical limit to respiration. With acute hypoxia, as ventilation increases, blood PCO_2 falls, resulting in respiratory alkalosis. If breathing increases there will be death from alkalosis, if breathing does not increase then death occurs due to hypoxia. Either way, the capacity for tolerance has been exceeded and death occurs.

With gradual exposure to high altitude there are additional adaptations. Long-term acclimatization produces increases in respiratory volume, blood volume and blood concentration. The principal adaptation is the increase in the concentration of blood cells in plasma. Arterial PO_2 (P_aO_2) is inversely proportional to erythropoietin secretion from the kidneys, and so low P_aO_2 leads to an increase in the number of red blood cells (polycythaemia). This is ultimately limited by corresponding increases in blood viscosity. Nevertheless, the oxygen carrying capacity of blood can be appreciably increased. Such adaptations over a period of weeks enable mountaineers to climb to almost 9 km (where PO_2 in

EFFECT OF ALTITUDE ON THE MIDDLE EAR:

*With an increase in altitude of 500 m or more, the air
in the middle ear expands sufficient to cause the eardrum to bulge taught.
This results in an uncomfortable sensation and attenuation of hearing acuity.
The pressure difference is equalized by either opening of the Eustachian tube
(facilitated by swallowing), or rupture of the eardrum.*

EFFECTS OF OXYGEN ON VASOMOTOR TONE:

*Hypoxia causes pulmonary vasoconstriction and systemic vasodilatation,
leading to pulmonary hypertension, cerebral oedema and increased urination.
Hyperoxia causes systemic vasoconstriction and pulmonary vasodilatation,
leading to insufficient cerebral blood flow and pulmonary oedema.*

air = 4 kPa) without oxygen, though they may only remain for a short time. The highest permanently inhabited places are at about 5.3 km (where PO_2 in air = 8 kPa), with regular work being carried out at 6.2 km (where PO_2 in air = 7 kPa) without any adverse effects.

Hyperventilation and increased urination are a part of normal acclimatization. There are three main types of altitude sickness, which can either progress from the most mild to the most severe, or one type may occur independently of the other two types. The most mild, acute mountain sickness, is characterised by headache, fatigue and nausea. Whereas high altitude cerebral oedema is characterised by confusion and ataxia. The most severe, high altitude pulmonary oedema, is characterised by breathlessness. Altitude sickness of any type usually occurs with a delayed onset of 6-12 hours post ascent, and the only cures are either acclimatization or descent.

Hyperoxia is an excessively high PO_2. Hyperoxia (oxygen toxicity or oxygen poisoning) occurs if PO_2 in air exceeds 40 kPa. Excess oxygen interferes with various metabolic processes and can actually inhibit respiratory oxidation. With babies kept in incubators, if the PO_2 exceeds 40 kPa, the vitreous body can become opaque causing irreversible blindness, due to retrolental fibroplasia (high oxygen causes vasoconstriction in the developing retina, subsequent withdrawal of oxygen then leads to an inappropriate and damaging growth of blood vessels within the eye). With adults on isobaric oxygen therapy, if the PO_2 exceeds 60 kPa, cardiac output decreases and pulmonary oedema can develop over days of exposure. Convulsions and unconsciousness occur rapidly if PO_2 exceeds 220 kPa (this is equivalent to diving to a depth of 100 m with compressed air).

SUMMARY OF OXYGENATION:

The respiratory and cardiovascular systems dominate oxygenation (and the complementary removal of carbon dioxide). The diaphragm and thoracic mechanics must move sufficient air in and out of the lungs, alveolar gas transfer must be unimpaired, whilst the blood must contain sufficient haemoglobin, and the heart must circulate the blood adequately. Oxygenation is directly affected by hydration, pH and temperature. In turn, oxygenation directly limits the ability of respiratory metabolism to yield energy within tissues.

FOR WRITTEN NOTES:

CHAPTER 5.

ENERGY

INTRODUCTION TO ENERGY:

Energy is the capacity for work. All active processes are therefore, by definition, entirely dependent upon energy. The derivation of energy requires the acquisition of substrates via nutrition and their modification via metabolism. Thus the gastro-intestinal tract is partly responsible for supplying the incessant demand for energy. Whilst metabolism is induced or inhibited via hormonal control. However, within all living organisms the release of energy is synonymous with the term respiration. Respiration is not just the mechanical process of breathing, but also encompasses the exchange and internal fate of gases common to all living cells. When considering energy the respiratory system predominates.

The most basic concept of a living thing is: an entity that extracts energy and materials from its surroundings, with which to maintain its composition and to synthesize others of the same kind. In the human body, there are many complicated active processes that are intricately interdependent for the continuation of life, but all rely first and foremost on energy. Energy expenditure can be either due to *A*nabolism during *B*iosynthesis, or *C*atabolism during *D*egradation. If the substrate utilized is either a fat or carbohydrate then the metabolism is said to be functional. If the substrate is a protein then the metabolism is structural. Thus energy is utilized to transform substances and these processes also convert the energy into different types of energy, such as mechanical, electrical, chemical or heat energy. The total energy conversion is equal to the sum of *external work, lost heat* and *stored energy*.

PHYSICS OF ENERGY:

Remember matter (whether living or not) is anything with both mass and volume. Relativity theory demonstrates that the energy and mass possessed by matter have to be considered as different aspects of the same thing. Neither can be created or destroyed, only converted; and within one body neither can be altered without affecting the other.

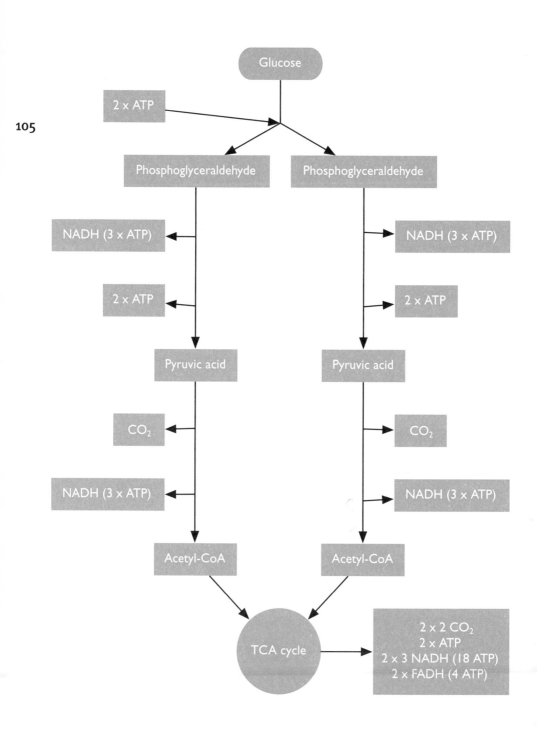

Einstein's law inspirationally links energy to the mass of a body:

$$E = m.c^2$$

Where; E = energy, m = mass and c = velocity of light (2.99792458×10^8 m.s^{-1} in a vacuum). The units of energy therefore being kg.m^2.s^{-2}.

The units of measurement for energy, heat and work are the same. One joule (J) can also be expressed as one N.m or one W.s (as well as one kg.m^2.s^{-2}); each of which is equivalent to 0.239 calories (as 1 cal is equal to 4.1868 J, averaged over 0-100 °C). Similarly, one kW.h equals 3.6 MJ, which is equivalent to 860 kcal. On the other hand, power is measured in watts (W). One watt is equal to one J.s^{-1} (as well as one kg.m^2.s^{-3}). Where, one kJ.h^{-1} is equal to 0.28 W, which is also equivalent to 0.239 kcal.h^{-1}. Similarly, one kJ.d^{-1} is equal to 0.012 W, which is also equivalent to 0.239 kcal.d^{-1}! One horsepower (HP) is equivalent to 736 W.

The efficiency of energy production and its use is closely tied to the three laws of thermodynamics (relating to the principles of energy, entropy and absolute zero respectively).

Where, % efficiency = external work / converted energy x 100.

Also, Net efficiency = Gross metabolism - basal metabolism.

Overall, the maximum efficiency for an isolated muscle is about 35 %, and for a whole organism it is 25 %. In comparison, the thermal efficiency of a steam piston engine is 7-20 %, for a steam turbine engine it is 25-30 % and for an internal combustion engine it is 30-37 %.

CHEMISTRY OF ENERGY IN THE BODY:

Adenosine Tri-Phosphate (ATP) molecules are the biological equivalent of a short-term battery. The conversion of energy from carbohydrate to ATP molecules is about 40 % efficient, with the remainder lost as heat. ATP stores energy until it is required for active processes, such as: transport, synthesis, nerve transmission or muscle contraction. Each cell needs in the order of two million ATP molecules each second

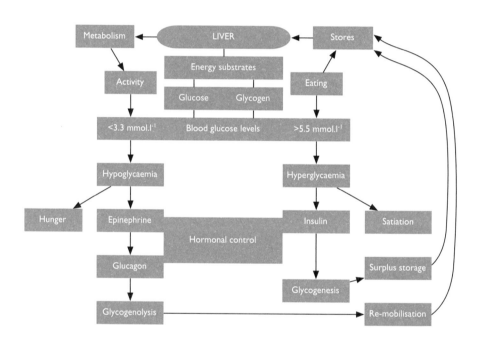

in order to drive its biochemical machinery. For ATP in muscle, this amounts to 5 $\mu g.g^{-1}$. There is typically a reserve of 5 $\mu mol.g^{-1}$ muscle, capable of powering ten contractions (1 mole of ATP being equivalent to about 30 kJ). Each ATP molecule consists of adenine and a ribose moiety onto which are attached three phosphate groups, two being joined by high-energy bonds. Hydrolysis of these bonds results in the release of energy. ATP is formed by the addition of inorganic phosphate (P_i) to Adenosine Di-Phosphate (ADP) via a high-energy bond, where:

ADP + P + 30 kJ = ATP + Water.

ATP is formed during aerobic cellular respiration, which has three catabolic stages. Stage one is the glycolysis of glucose in the cytoplasm to produce two molecules of pyruvic acid. Overall, glycolysis consumes two and produces ten ATP molecules. Stage two is the oxidation and decarboxylation of pyruvic acid in mitochondria to produce acetylcoenzyme A (acetyl-CoA). Each pyruvic acid molecule yields three ATP molecules. Stage three is the tricarboxylic acid cycle (TCA, citric acid or Kreb's cycle) in mitochondria. Each acetyl-CoA molecule results in one turn of the TCA cycle and the production of twelve ATP molecules. Therefore each glucose molecule requires two turns of the TCA cycle and produces a net of 38 ATP molecules.

ROLE OF THE LIVER:

The structure and function of the liver is very important in regulating the supply of energy to meet demands. The most important source of energy is carbohydrate, particularly glucose. Blood sugar levels are equal to the balance of intake and synthesis against utilization. Serum glucose should be between 3.3 and 5.5 $mmol.l^{-1}$ (when fasting, or up to 10 $mmol.l^{-1}$ 90 minutes after eating). These blood sugar levels are under hormonal control. The principal hormones involved are insulin, glucagon and epinephrine (previously known as adrenaline). The role of the liver in energy balance is to control metabolism and storage, based on blood sugar levels. Blood glucose rises after eating, causing the secretion of insulin from the pancreas, which in turn results in glycogen*esis* for storage of the surplus. The hyperglycaemia also causes satiation. Conversely, blood glucose falls during activity, causing the secretion of epinephrine from the adrenal medulla, which in turn leads to the secretion of glucagon from the pancreas, and results in glycogen*olysis*

109

for the re-mobilization of stored reserves. The hypoglycaemia also causes hunger.

Insulin imbalance is a common problem. An excess of insulin leads to hypoglycaemia, and can be caused by alcohol or drugs. Less than 2 mmol.l^{-1} (equivalent to less than 0.35 g.l^{-1}) of glucose in blood results in cerebral impairment. An inadequate insulin response leads to hyperglycaemia, and can be caused either by the pancreas secreting insuffienct insulin (type 1, juvenile-onset or insulin-dependent diabetes mellitus), or resistance to high levels of circulating insulin (type 2, adult-onset or non-insulin-dependent diabetes mellitus). Glucose is lost via sweet urine (glucosuria or glycosuria). As less glucose is stored, there is a greater dependence on fat metabolism for energy. The metabolism of fats results in an increase in fatty acids and thus metabolic acidosis. Acidosis is spread throughout the body by the blood, and progressively affects respiration, the kidneys and the liver. If acidosis continues, the resultant multiple organ failure (MOF) eventually causes death.

METABOLIC RATE:

The basal metabolic rate (BMR) is the power of a body, measured while at rest, in the morning, after fasting and under thermoneutral conditions. The basal metabolic rate of a 70 kg body is 6-7 MJ.d^{-1}. This is equivalent to about 4 kJ.kg^{-1}.h^{-1} or approximately 1 W.kg^{-1} (cf. one small chocolate bar contains of the order of 500 kJ). Within the normal range, women have the lowest and men have the highest basal metabolic rates. This is because women have more fat deposits than men, and the insulation this extra fat provides, means that a lower metabolic rate is needed to maintain body temperature at rest. The metabolic rate can increase to 20 MJ.d^{-1} (12 kJ.hour^{-1}) during moderate activity, and up to 100 MJ.d^{-1} (1 kJ.min^{-1}) during extreme physical exercise. Metabolic rate also increases during mental activity, but only due to increases in muscle tone. The actual metabolic demands of the brain remain fairly constant irrespective of the activity, even during sleep. Basal metabolic rate also follows a diurnal rhythm, with an increase in the morning and a decrease at night. Comparison of the lowest base level thus requires measurements to be made first thing in the morning. Furthermore, the consumption of energy yielding substrates increases metabolic rate by about 6 % (cf. 6 % of ingested energy is not absorbed by the gastro-intestinal tract; equivalent to 820 kJ.d^{-1}). This elevation of metabolic

FOR WRITTEN NOTES:

111

rate by eating is called diet-induced thermogenesis, and is greater for protein and less for carbohydrate or fat. Finally, thermoregulatory responses involve the expenditure of energy and so will increase the metabolic rate even when resting in the supine position.

It is possible to use the analogy of a light-bulb to convey how much power is utilized by a living human. The typical basal metabolic rate approximately equates to between one and four 60 W light-bulbs (or up to twenty 60 W light-bulbs during exercise). At rest one-half of this energy is lost from the body as heat. (Total combustion is the physical caloric, whereas partial combustion is the physiological caloric value.) The body can utilize different energy substrates, although with different efficiency. Complete oxidation of fat yields 38.9 kJ.g^{-1}, carbohydrate yields 17.2 kJ.g^{-1} (glucose yields 15.7 kJ.g^{-1}), and protein (egg albumin) yields 24 kJ.g^{-1}. It is interesting to note that ethyl-alcohol yields 29.7 kJ.g^{-1}. The energy yielded is isodynamic, meaning that it is interchangeable in use regardless of source.

The relative contribution of structures to the basal metabolic rate is: 26 % from liver, 26 % from muscle, 18 % from brain, 9 % from heart, 7 % from kidney and 14 % from the remainder of the body. During sleep or undernutrition it is possible for metabolism to fall below the basal rate. With sleep there is loss of muscle tone, and hence a decrease in the contribution to basal metabolic rate due to muscle. Whereas with undernutrition there is a decrease in liver activity, and hence a decrease in the contribution to basal metabolic rate due to the liver. Overall, at rest about 55 % of the energy generated by metabolism is released as heat. Most of this heat is from the muscles (33 %) and the liver (13 %). However, heat is distributed throughout the body by transport in the blood. Furthermore, heat loss depends on the temperature gradient between the internal source and the external surface. Therefore it is not uncommon for 40-50 % of body heat to be lost through the head, principally because the brain is highly vascularized and poorly insulated.

RESPIRATORY QUOTIENT:

The respiratory quotient (RQ) is an indicator of the metabolic substrate utilized to derive energy. The RQ depends on the balance of oxygen and carbon dioxide respired (RQ refers to the number of moles in tissue, whereas respiratory exchange ratio refers to the volumes respired),

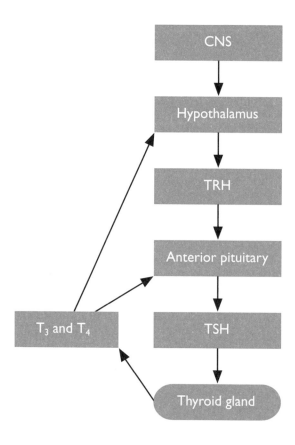

where: RQ = CO_2 produced / O_2 consumed.

In the oxidation of glucose ($C_6H_{12}O_6$), the amount of carbon dioxide produced is the same as the amount of oxygen consumed.

$C_6H_{12}O_6 + 6O_2$ reacts to give $6CO_2 + 6H_2O$ + 2.826 MJ (via ATP)

Therefore an RQ of 1 indicates carbohydrate oxidation. It can be calculated that glucose yields 15.7 $kJ.g^{-1}$ (cf. 1 mole of glucose = 180 g, and 6 moles of oxygen = 6 x 22.4 l = 134.4 l, therefore 1 g of glucose yields 2,826 / 180 = 15.7 $kJ.g^{-1}$). At an RQ of 1, oxygen yields 21 $kJ.l^{-1}$ (cf. 2,826 / 134.4 = 21 $kJ.l^{-1}$). The oxidation of fat gives an RQ of 0.7 and oxygen yield of 19.6 $kJ.l^{-1}$. The oxidation of protein gives an RQ of 0.81 and oxygen yield of 18.8 $kJ.l^{-1}$. Thus, under normal circumstances the RQ varies from 0.7 to 1. On a mixed diet, the average RQ is 0.82, giving an oxygen yield of 20.2 $kJ.l^{-1}$. It can be seen that every change in RQ of 0.1 corresponds to an equivalent change in oxygen energy yield of about 1 $kJ.l^{-1}$. Thus glucose yields the most energy per unit of oxygen.

It is possible for the RQ to exceed the limits that would be indicated by the straightforward metabolism of the different substrates. Hyperventilation or excess carbohydrate intake can increase RQ to 1.4. With hyperventilation, oxygen uptake cannot be increased as it is already maximal, but additional carbon dioxide is lost from the buffer reserves in blood. Thus a higher proportion of carbon dioxide is exhaled through the lungs. With overfeeding on carbohydrates, carbohydrate is converted into fat. As fats contain less oxygen than carbohydrates, the conversion from carbohydrate to fat effectively releases oxygen. Thus a lower proportion of oxygen is taken up from the lungs. At the other extreme, fasting or diabetes can decrease RQ to 0.6. This is due to the additional oxygen required for the remobilization of the energy stored in fat. Thus a higher proportion of oxygen is taken up from the lungs.

HYPOTHYROIDISM AND HYPERTHYROIDISM:

The basal metabolic rate is primarily controlled through hormones secreted by the thyroid gland. Abnormal levels of thyroid hormone are usually associated with the development of an enlarged thyroid gland (goitre). The hypothalamus secretes thyrotrophin-releasing hormone (TRH), which in turn stimulates the pituitary gland to secrete

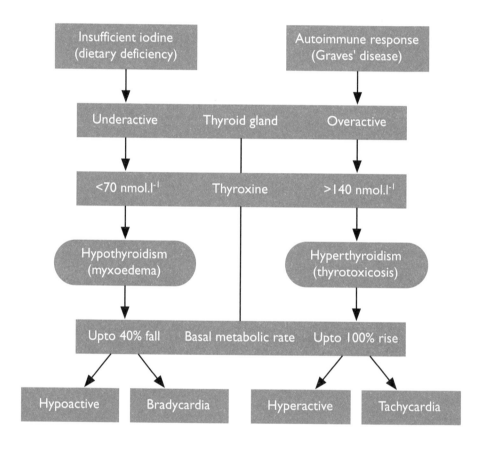

thyroid-stimulating hormone (TSH, thyrotrophin or thyrotrophic hormone). TSH increases the secretion of triiodothyronine (T_3) and thyroxine (T_4) from the thyroid gland. In turn, T_3 and T_4 decrease the secretion of TSH. Most T_3 and T_4 in plasma is bound to thyroxine-binding globulin (TBG). It is the unbound T_3 and T_4 that are active, and are responsible for increasing the rate of metabolism and catecholamine effects. T_3 is five-fold more active than T_4. However, the thyroid gland produces mainly T_4, and 85 % of T_3 is produced from the conversion of T_4 in blood and other tissues. Thus it is predominantly changes in the production of T_4 by the thyroid gland that regulate the metabolic rate.

When the thyroid gland is under-active, insufficient thyroxine is produced leading to hypothyroidism (which can result in myxoedema). The level of thyroxine in blood decreases below 70 nmol.l^{-1}, and basal metabolic rate can decrease by 40 %. There is an increase in weight, decrease in heart rate (bradycardia), lethargy, dry skin, thin hair, and (because of difficulty in increasing metabolic rate to increase heat production) a dislike of cold weather. This is classically caused by a deficiency of iodine in the diet.

When the thyroid gland is overactive, excess thyroxine is produced leading to hyperthyroidism (which can result in thyrotoxicosis). The level of thyroxine in blood increases above 140 nmol.l^{-1}, and basal metabolic rate can increase by 100 %. There is a decrease in weight, increase in heart rate (tachycardia; even when sleeping), hyperactivity, increase in hunger, sweating, and (because of a high metabolic rate increasing heat production) a dislike of hot weather. This is classically caused by an autoimmune condition called Graves' disease.

SUMMARY OF ENERGY:

Energy is the fundamental basis of all life. Metabolism encompasses all the chemical reactions by which energy is manipulated within the body. The most efficient means of yielding energy is respiratory metabolism. This requires the utilization of water as both a medium and product, while consuming oxygen, decreasing pH and increasing temperature. The principal side effect of metabolism is heat generation. The production of heat provides for thermoregulation. Thermoregulation is needed in order to optimize chemical reactions, and in turn drives a continuously high metabolic rate.

For written notes:

117

CHAPTER 6.

ACIDITY

INTRODUCTION TO pH:

The potential of hydrogen (pH) is a measure of the hydrogen ion concentration ($[H^+]$) in an aqueous solution.

$$pH = log_{10}.(1 / [H^+])$$
$$= - log_{10}.[H^+]$$

The higher the number of hydrogen ions the lower the pH value. Possible values range from 1.0 (highly acidic; $[H^+] = 10^{-1}$ mol.l^{-1}) to 14.0 (highly alkaline; $[H^+] = 10^{-14}$ mol.l^{-1}), the mid-point at 7.0 represents a neutral pH ($[H^+] = 10^{-7}$ mol.l^{-1}). As the scale is logarithmic, each change of 1.0 in either direction equates to a ten-fold change in the number of hydrogen ions. The scale is used as the electrochemical potential of ions is not directly proportional to their concentration, but rather to the logarithm of their concentration. Thus the acid-base responses of living processes tend not to be directly proportional to $[H^+]$, but rather to pH.

The word acid conjures in our minds the vision of something strong, powerful and dangerous. It is true that extreme acidity has profound implications for living processes, but so does extreme alkalinity.
The internal environment of the living body must be regulated to provide optimal conditions for the chemical reactions that are required. In particular the derivation of energy via metabolism must be efficient in order for an organism to compete and survive. The production and consumption of hydrogen ions (pH) is of fundamental importance.
For example, in aerobic respiration the metabolism of each glucose molecule produces 24 hydrogen ions. Overall, it is the cardiovascular, respiratory and renal systems that work together to maintain a constant acid-base balance.

CHEMISTRY OF pH IN THE BODY:

The respiratory system is responsible for the transport and exchange of gases. This serves to ensure that all tissues receive the oxygen they need, and can dispose of all the carbon dioxide they produce.

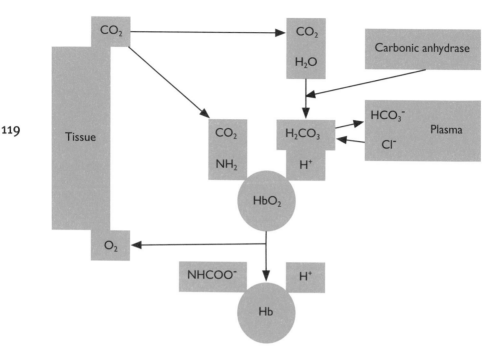

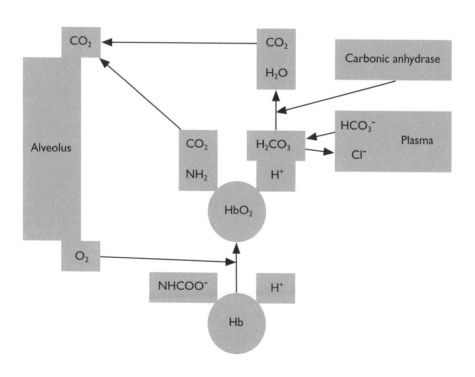

Carbon dioxide is more soluble than oxygen and also reacts chemically with water. Surprisingly there is much more carbon dioxide in blood (almost three times as much carbon dioxide in arterial blood) than oxygen, both more dissolved and more reacted chemically. This is because carbon dioxide is a major part of the system controlling blood pH. The role of carbon dioxide in acid-base balance is as important a process as the transport of carbon dioxide from tissues to the lungs. Therefore first consider carbon dioxide in *arterial* blood, as this has a relatively constant carbon dioxide level. At a PCO$_2$ of 5.3 kPa, water dissolves 1.2 mmol.l^{-1}. In plasma, dissolved carbon dioxide reacts with water forming hydrogen and bicarbonate ions, via the intermediary carbonic acid:

$$CO_2 + H_2O \Leftrightarrow H_2CO_3 \Leftrightarrow H^+ + HCO_3^-$$

In erythrocytes, this reaction is 10,000 fold faster because of the presence of the enzyme carbonic anhydrase. The reaction is reversible and the amount reacting depends on the concentrations of reactants and products. The amount of dissolved carbon dioxide depends directly on PCO$_2$. Changes in P$_a$CO$_2$ will affect pH, because the arterial blood is normally in equilibrium with a [H$^+$] of 40 nmol.l^{-1}, which is about one millionth of the [HCO$_3^-$] of 24 mmol.l^{-1}. So if equal amounts of hydrogen and bicarbonate ions are added or removed, by increasing or decreasing P$_a$CO$_2$ and shifting the equilibrium to the right or left, then the change in [H$^+$] will be disproportionately larger than the change in [HCO$_3^-$]. So if P$_a$CO$_2$ rises pH will fall, and if P$_a$CO$_2$ falls pH will rise. However, not only does most of the reaction occur in erythrocytes, but also within erythrocytes, one of the products (H$^+$) is removed. Hydrogen ions bind to haemoglobin. When this occurs, more carbon dioxide reacts and more bicarbonate is formed (as can be seen from the equation above). Bicarbonate leaves the red cell in exchange for inward movement of chloride ions (chloride shift, or Hamburger's shift when reversed), maintaining about 24 mmol.l^{-1} of bicarbonate ions in plasma. Plasma bicarbonate does not usually change much with changing P$_a$CO$_2$, because the reactions of carbon dioxide in the red cell are mostly determined by how much H$^+$ binds to Hb.

So the pH of plasma primarily depends on the ratio between the reaction of carbon dioxide in the red cell, and the reaction of carbon dioxide in the plasma. In the whole body, respiration controls the P$_a$CO$_2$ by varying

CO$_2$ IN ARTERIAL BLOOD		%	MMOL.L^{-1}
Plasma (70%)	CO$_2$	3	1.2
	HCO$_3^-$	65	24
	Carbamino-plasma proteins	2	0.8
Erythrocytes (30%)	CO$_2$	3	1.2
	HCO$_3^-$	20	8
	Carbamino-haemoglobin	7	2.8
Total		100	38

All are slightly elevated in venous blood,
such that an additional 2-4 mmol.l^{-1} of carbon dioxide is carried;
only 5-10 % of the total.

Of the carbon dioxide transported:
60-85 % travels as bicarbonate,
9-30 % travels bound to protein,
6-10 % travels as dissolved carbon dioxide.

ventilation (230 ml.min^{-1} carbon dioxide corresponds to about 15 mol.d^{-1}). Meanwhile, the kidney controls the bicarbonate concentration in plasma by varying excretion. (NB the kidney also excretes 40-60 mmol.d^{-1} H$^+$ by binding to HPO_4^{2-} and NH_3.) Overall, the control of pH is primarily attributed to the lungs (for respiratory control of P_aCO_2) and the kidneys (for metabolic control of $[HCO_3^-]$). This is described by the Henderson-Hasselbalch equation:

$$pH = pK' + \log.([HCO_3^-] / P_aCO_2 \times \alpha)$$

Where; pK' = dissociation constant for carbonic acid in plasma = 6.1, α = solubility coefficient = 0.225 mmol.l^{-1}.kPa^{-1}.

CO_2 also binds directly to protein such as haemoglobin, to form carbamino compounds:

$$Hb-NH_2 + CO_2 \Rightarrow Hb-NHCO_2^- + H^+$$

Blood pH is also buffered by the binding of hydrogen ions with proteins to form proteinate. This is particularly true of albumin in plasma and haemoglobin in erythrocytes. Slightly more is bound in venous blood because carbon dioxide binds more strongly to deoxygenated haemoglobin (the Haldane effect), helping to remove carbon dioxide from tissues. Furthermore, deoxyhaemoglobin is less acidic than oxyhaemoglobin. Thus, haemoglobin can directly contribute to buffering the acidity of carbon dioxide after it is exchanged for oxygen. With increasing P_aCO_2, more bicarbonate and hydrogen ions are produced. These hydrogen ions are then bound in proteinate. Thus with increasing P_aCO_2, there is a decrease in the concentration of proteins available for buffering. This is matched by an increase in the concentration of bicarbonate available for buffering. The reverse is true if P_aCO_2 falls. The total concentration of buffer bases includes both proteinate and bicarbonate, and is therefore constant irrespective of P_aCO_2. The total concentration of buffer bases is 48 mmol.l^{-1}. More than 48 mmol.l^{-1} reflects a base excess (BE), whereas less than 48 mmol.l^{-1} reflects a base deficit.

CLINICAL MEASUREMENT OF pH:

Clinically, blood-gas analysis is performed to gain information regarding

123

Clinical parameters assessed and their normal values:

pH = 7.35-7.45

P_aO_2 = 10.6-13.3 kPa

P_aCO_2 = 4.7-6.0 kPa

Base excess = +2 to -2 mmol.l^{-1}

Standard bicarbonate = 22-26 mmol.l^{-1}

respiratory and metabolic status. A small (0.5 ml) blood specimen is taken from an artery, usually the radial artery if it is patent, and analysed. First of all, the pH gives an indication of overall blood acidity. A change in pH of 0.3 represents a two-fold change in $[H^+]$. Secondly, the P_aO_2 reflects diffusion from alveolus to erythrocyte (assuming adequate haemoglobin). If P_aO_2 is low and P_aCO_2 is high then there is insufficient ventilation. If P_aO_2 is low and P_aCO_2 is low or normal, then there is either a diffusion defect (e.g. interstitial oedema) or ventilation-perfusion imbalance (e.g. shunt). Thus a respiratory problem can be detected, which may help explain an acid-base imbalance. Thirdly, the P_aCO_2 gives an indication of respiratory state, irrespective of pH. A high P_aCO_2 reflects respiratory acidosis, whereas a low P_aCO_2 reflects a respiratory alkalosis. Finally, the base excess gives an indication of metabolic state, also irrespective of pH. A negative BE reflects metabolic acidosis, whereas an excessively positive BE reflects metabolic alkalosis. (NB an approximate BE can be estimated by taking away 24 from the standard bicarbonate, although this does not take account of potential proteinate effects.) Arterial blood-gas measurements provide a crucial indicator of prognosis: "No one dies with normal blood gases."

Normal Body pH:

pH varies depending on the part of the body concerned and the time it is measured. The greatest range is found in the gastro-intestinal tract, where there are secretions in response to digestive requirements. There, gastric secretions can be as low as 1.1 and pancreatic secretions as high as 8.8. A diet of mostly animal protein has a slightly acidic pH, and leads to putrefaction predominating in the gastro-intestinal tract. A diet of mostly plant carbohydrate has a slightly alkali pH, and leads to fermentation predominating in the gastro-intestinal tract. Thus, a balanced diet should not contribute a change in the pH of the body. Normally, intracellular pH is 6.8-7.2, whereas extracellular pH is close to 7.4. Plasma pH is usually carefully maintained between 7.35 and 7.45 (equivalent to $[H^+]$ of about 45-35 $nmol.l^{-1}$ respectively).

Changes in pH Balance:

If blood becomes too acid or alkaline there are serious consequences. If plasma pH falls below 7.0, enzymes are lethally denatured.

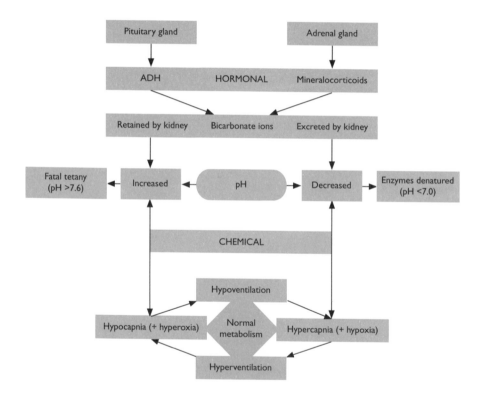

Conversely, if plasma pH rises above 7.6, free calcium concentration falls (as H^+ falls, Ca^{2+} is sequestered as a substitute for H^+) enough to produce fatal tetany (tetany is muscle cramps and spasms due to a decreased activation threshold for action potentials, thus there is an increase in excitation and hence involuntary muscular contractions).

pH is controlled by chemical buffering reactions and hormonal effects, on both water and ionic balance within body fluids. Buffering depends on the reactions in blood and in turn their control of breathing and urine output. For example, buffering can occur if the body produces acid. The acid reacts with bicarbonate to form carbon dioxide, which is breathed out, stopping the pH from falling. However, arterial PO_2 and PCO_2 need to be kept constant. P_aO_2 should be >10.6 kPa, and P_aCO_2 should be between 4.7 and 6.0 kPa. If ventilation increases with no change in metabolism (hyperventilation), P_aCO_2 will fall and P_aO_2 will rise. Whereas if ventilation decreases with no change in metabolism (hypoventilation), P_aCO_2 will rise and P_aO_2 will fall. However it is not always possible to control both partial pressures by changing the ventilation rate. If P_aO_2 falls and P_aCO_2 rises, then it is possible to correct both by breathing more. But, if PO_2 falls with no change in PCO_2 (as occurs at high altitude), correcting the hypoxia will produce a hypocapnia (a low P_aCO_2). Sometimes the system can not rectify all imbalances. In addition to this chemical control of breathing, there are hormonal influences on urinary output. Particular hormones that influence pH are the mineralocorticoids and ADH. Mineralocorticoids from the zona glomerulosa in the cortex of the adrenal gland facilitate the retention of sodium, bicarbonate and chloride ions in the kidney. Similarly, ADH from the pars nervosa of the pituitary gland prevents excess loss of water, sodium ions and co-transported hydrogen ions in urine. Urine is normally acidic with a pH of 6.0, and so usually contains hydrogen ions but no bicarbonate ions. However bicarbonate ions can be excreted in urine, for example if needed in order to compensate for alkalosis. Whereas renal excretion of hydrogen ions can be either increased or decreased. Thus urinary pH can range from 4.5-8.0.

Finally, cerebro-spinal fluid (CSF) pH is determined by the ratio of $[HCO_3^-]$ to PCO_2. $[HCO_3^-]$ is fixed in the short-term. So falls in PCO_2 lead to rises in CSF pH, and rises in PCO_2 lead to falls in CSF pH. However, persistent changes in pH are corrected by choroid plexus cells changing the $[HCO_3^-]$, by altering the transport of HCO_3^- from the CSF. Thus over time

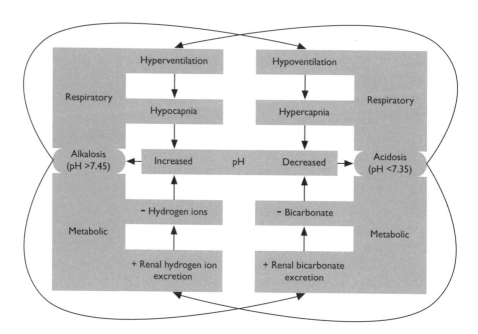

a normal CSF pH can be maintained, with either a sustained hypocapnia due to hyperventilation (e.g. at high altitude), or sustained hypercapnia due to hypoventilation (e.g. caused by chronic lung disease).

ACIDAEMIA AND ALKALAEMIA:

Acidaemia reflects an excess [H+]. With acidaemia blood pH falls below 7.35, brought about by an acidotic process. Conversely, alkalaemia reflects an insufficient [H+]. With alkalaemia blood pH rises above 7.45, brought about by an alkalotic process. In both cases, the process that is primarily responsible can be either respiratory or metabolic. Hypoventilation leads to hypercapnia, and hypercapnia causes plasma pH to fall. This is *respiratory acidosis*. On the other hand, hyperventilation causes P_aCO_2 to fall and so pH rises. This is *respiratory alkalosis*. However, small changes can be compensated by the kidney increasing [HCO_3^-] or [H+] in urine. Plasma pH depends on the ratio of [HCO_3^-] to P_aCO_2, not on their absolute values. Changes in P_aCO_2 can be compensated by changes in [HCO_3^-] and [H+]. The kidney decreases [HCO_3^-] or [H+] in blood by variable excretion. Renal changes occur over hours to days, whereas respiratory changes occur more rapidly over seconds to minutes. If tissue metabolism produces *acid*, this reacts with bicarbonate ions. The fall in [HCO_3^-] leads to a fall in pH. This is *metabolic acidosis*, and can be compensated for by changing ventilation. Increased ventilation lowers P_aCO_2, restoring pH towards normal. Conversely, if plasma [HCO_3^-] rises (e.g. due to loss of hydrogen ions after vomiting) plasma pH rises. This is *metabolic alkalosis*, and can also be compensated for (to a degree) by changing ventilation. Decreased ventilation raises P_aCO_2, restoring pH towards normal.

SUMMARY OF PH:

pH depends primarily on the buffering properties of the blood, (particularly the utilization of bicarbonate and proteinate), combined with the control of carbon dioxide elimination by respiration, and bicarbonate and hydrogen ion excretion by the kidneys. These mechanisms combine to enable the correction of systemic acid-base imbalances. Changes in pH affect both the structure and function of proteins by removing them from their optimum iso-electric point. Hence optimal pH is essential to maximize the efficiency of energy yielding respiratory metabolism.

For written notes:

129

CHAPTER 7.

TEMPERATURE

INTRODUCTION TO TEMPERATURE:

Temperature is a measure of the degree of hotness (or coldness) of a substance, medium or body. Thermodynamics is derived from the Greek terms therme and dynos, meaning heat and power respectively. Thus thermodynamics is the physical science concerned with the relationships between heat and mechanical work. Primarily it is the first two laws of thermodynamics that concern us here. The first law states that when one form of energy is converted into another form of energy there is no loss or gain. The second law states that when one form of energy is converted into another form of energy a proportion is also converted into heat. With respect to life, thermoregulation is the control of body heat in homoiothermic (homeothermic, endothermic or warm-blooded) organisms. The key processes are therefore movement and the derivation of energy via metabolism. The musculoskeletal, cardiovascular and respiratory systems all contribute to both movement and metabolism in their different ways.

PHYSICS OF TEMPERATURE:

Units:
The unit of absolute temperature is the Kelvin (K), equal to the fraction of 1/273.16 of the thermodynamic temperature (T) of the triple point of water. The triple point of water is equal to the temperature at which ice, liquid water and water vapour coexist in equilibrium (and is 273.16 K or 0.01 °C). Temperature was originally based on the unit of a degree, where one degree was derived as one-hundredth of a division between the temperature at which pure ice melts (273.15 K or 0 °C) and that at which pure water boils (373.15 K or 100 °C) in water-saturated air at 101.325 kPa (atmospheric pressure at sea level). The Celsius (previously known as Centigrade) scale of temperature was thus based on the behaviour of water. Now the Celsius temperature (°C) is equal to $T - T_0$, where T_0 is 273.15 K. Temperature conditions can be: 'Standard' at 0°C (273.15 K), 'Body' at 37 °C (or about 310 K), or 'Ambient' at environmental or room temperature (sometimes taken as 25 °C).

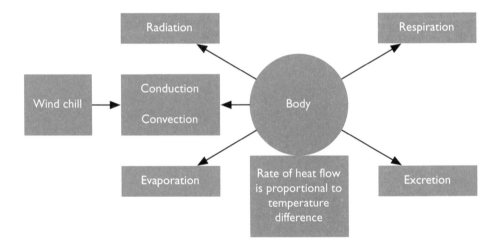

Heat Transfer:
Heat is the energy transferred as a result of a difference in temperature. Heat transfer mechanisms are obviously of crucial importance in the regulation of body temperature. For heat to flow there must be a difference in temperature, and then heat will only flow from hot to cold. The rate of heat transfer will be proportional to the difference in temperature. From physics we know that radiation, conduction, convection and evaporation are the mechanisms of heat transfer. With biology we must also remember to consider respiration and excretion.

Radiation refers to the total energy radiated per area:

Radiation, $R = \sigma . (T_s^4 - T^4)$ in $W.m^{-2}$

Where; Stefan's constant, $\sigma = 5.7 \times 10^{-8}$ $W.m^{-2}.K^{-4}$, T_s = skin temperature and T = environmental temperature (both in K).

Conduction and convection refer to the total energy conducted per unit area:

Conduction, $C = k . (T_2 - T_1) / d$ in $W.m^{-2}$

Where; k = thermal conductivity of medium in $W.m^{-1}.K^{-1}$ (0.6 for water, 0.046 for fat, 0.042 for skin and 0.025 for air; NB water has a 24-fold greater conductivity than air), $(T_2 - T_1) / d$ = temperature gradient. This gradient is the difference in temperature (in K) over a distance (in m). NB Conduction and convection will also be affected by any 'wind chill'.

Evaporation refers to the total energy lost by evaporation:

Evaporation, $E = m.L$ in kJ (kW.s)

Where; specific latent heat of evaporation (vaporization), $L = 2425$ $kJ.kg^{-1}$ (at 33 °C, for skin surface), m = kg of water.

Heat transfer from the Body:
In environmental conditions of extreme heat, up to 1.6 $l.h^{-1}$ of sweat can be produced. Where water is secreted out onto the surface of the skin, each litre that evaporates causes a loss of 2425 kJ. If the air is dry, evaporation can be used to maintain a normal body temperature up to

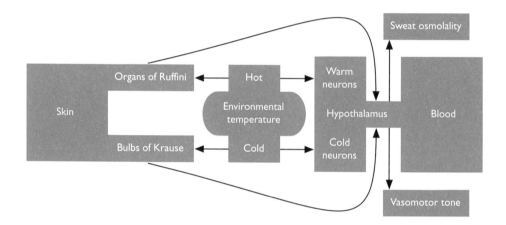

an environmental temperature of 100 °C. However, if the air has 100 % humidity then evaporation is ineffective and a normal body temperature can not be maintained above an environmental temperature of 33 °C.

Body temperature depends on both the heat produced and lost by the body. Total heat loss varies with environmental temperatures:
at 20 °C, 63 J.m^{-2}.s^{-1} is lost (E = 13 %, C = 26 %, R = 61 %),
at 30 °C, 38 J.m^{-2}.s^{-1} is lost (E = 27 %, C = 27 %, R = 46 %), and
at 36 °C, 43 J.m^{-2}.s^{-1} is lost (E = 100 %, C = 0 %, R = 0 %).

DETECTION OF TEMPERATURE BY THE BODY:

Temperature changes can be detected both externally and internally by thermoreceptors. Changes in environmental temperature are detected by specialized nerve-endings in the skin: the organs of Ruffini for warmth and the smaller bulbs of Krause for cold. The cutaneous cold receptors are more numerous and more evenly distributed than the cutaneous warm receptors. Changes in internal temperature are detected in blood by the hypothalamus. The hypothalamus has a resting temperature of 38 °C, and changes of up to 2 °C cause a two-fold change in the discharge rate of certain cells. There are two types of thermosensitive ganglion cells. The majority of thermosensitive cells respond only to increases in temperature and are called 'warm' neurons, the remainder respond to decreases in temperature and are called 'cold' neurons. Both cell types are found in the preoptic and anterior regions of the hypothalamus, with the posterior region acting as an integrative centre. To a lesser degree, the midbrain, medulla oblongata and spinal cord also demonstrate temperature sensitivity (and are also crucial for relaying some thermoregulatory signals from higher up). However, it is the hypothalamus that has an overriding influence on involuntary thermoregulatory responses. For example, the hypothalamus controls both the degree of vasoconstriction in response to changes in temperature, and the degree of ADH secreted in response to the changes in osmolality associated with sweating.

External and internal receptors act together. When external temperature changes, initially external and then internal receptors are stimulated. Similarly when internal temperature changes, initially internal and then external receptors are stimulated. Warm and cold receptors obviously act antagonistically, but external and internal receptors can

TEMPERATURE

135

also act antagonistically. For example, in order to protect against cold, a decrease in external temperature will cause an increase in the activity of cutaneous cold receptors, stimulating mechanisms that lead to a rise in core temperature. To prevent overcompensation, an increase in internal temperature will cause an increase in the activity of warm neurons, stimulating mechanisms that lead to a fall in core temperature. Overall, cold receptors predominate at the exterior, while warm receptors predominate in the interior. This is what we would expect if the body core is usually hotter than the environment. Thermoregulation is predominantly concerned with detecting low external temperature (as the environment is usually below optimum body temperature), and detecting high body core temperature (as any increase above basal metabolic rate associated with varying activity will produce more heat).

CLINICAL MEASUREMENT OF TEMPERATURE:

Measurement of body temperature can be made either externally or internally. The typical external sites are sublingual (beneath the tongue), oesophageal (through the mouth), otological (in the external auditory meatus), axillary (in the armpit), or rectal (through the anus; to a depth of 8 cm in an adult). Internal measurements are in direct contact with the interior of the body, normally the circulation via a needle inserted into a blood vessel. Measurements are typically made using either a liquid thermometer or a transducer. It is also possible to compare temperature measurements at different points in the body by thermography, i.e. using an infrared camera to identify 'hot spots' by localized differences in body temperature. This relies on the distribution of heat being proportional to the degree of vascularization. There are many important applications of thermography, such as: detection of tumours, investigation of bone fractures, placental localization, mapping of blood vessels (there are 9.7×10^7 m of vasculature in the body), and identifying the extent of gangrene, burns and frostbite. Regardless of the method of temperature measurement, it is always necessary to take account of insulation between the body and the measuring device, and to allow time for temperature to adjust before recording a value.

For written notes:

Normal Body Temperature:

Humans are homeothermic, with a core temperature of about
37+/-0.5 °C. It is impossible to define exactly what constitutes the body
core; generally it consists of the head and trunk. As heat is transferred
from the core to the surface there are temperature gradients throughout
the body. These gradients are complex because different body parts
vary in shape, circulation and insulation. Under normal circumstances a
10 °C difference between the core and extremities can occur. The mean
skin temperature is normally 33-34 °C when averaged over the whole
body. Even measurements at external sites supposedly representative
of the core temperature, can be expected to vary by up to 1 °C from core
temperature. Indeed even the core temperature itself can vary by up to
1 °C, for example from the hypothalamus near the centre of the brain to
the surface of the cortex. When making measurements it is important
to take account of these differences. The optimum body temperature
('set point') depends on the point of the body concerned. In addition,
for every point of the body there is a broad constellation of different
threshold temperatures associated with the activation of various
thermoregulatory mechanisms. At rest, shivering, sweating and blood
flow through the skin all correlate closely to the difference between
internal core temperature and the external skin surface temperature.

Body temperature also follows a pattern of cyclical fluctuations with
time. There is a diurnal rhythm with a minimum at c. 3-6 am and a
maximum at c. 3-6 pm, with a total difference of 0.5-1.5 °C. This rhythm
is inverted in nocturnal mammals, suggesting a correlation with physical
activity. However, night shift workers retain the same temperature
rhythm as day shift workers. Thus at night, workers have a slightly
lower body temperature and have to increase their metabolic rate a little
further to achieve the same level of activity, and hence are less efficient
(in a cold environment). To maximize efficiency in the long-term, diurnal
fluctuations in body temperature would be expected to reflect diurnal
fluctuations in physical activity. In fertile women there is also a longer
rhythm superimposed on the diurnal temperature rhythm. This concerns
the menstrual cycle, and involves a 0.2-0.5 °C rise after ovulation. The
temperature increase is thought to facilitate optimal implantation of the
ovum. Considering the other end of life, upon death body temperature
falls by 2 °C in the first two hours and then falls by 1.5 °C for each hour
thereafter, until reaching equilibrium with environmental temperature.

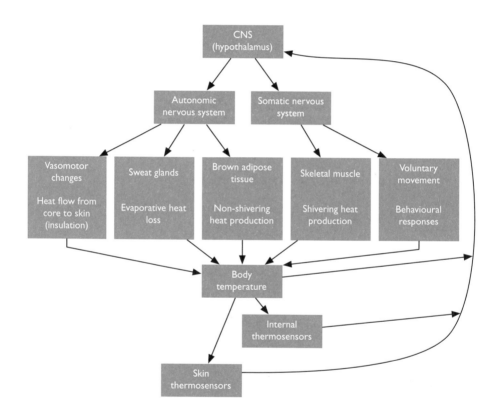

THERMOREGULATORY RESPONSES:

For thermoregulation the production and removal of heat must be equal. Changes in the rates of production, conservation and removal can be either voluntary (behavioural) or involuntary (autonomic). Involuntary production is by rhythmic muscular contraction (shivering), tonic muscular contraction, or increasing metabolism (non-shivering thermogenesis). Non-shivering thermogenesis is facilitated by brown adipose tissue, which is found predominantly between the shoulder blades and in the armpit (axilla) of babies. Brown adipose tissue is rich in mitochondria, and can cause a 2-5 fold increase in the basal metabolic rate of the body. This is also more efficient than shivering, as there are no rhythmic contractions that promote air currents and the associated increase in convective heat loss. Non-shivering thermogenesis is particularly important for babies. This is because babies have a surface area to volume ratio that is three-fold larger than adults, in addition to obvious restrictions on behavioural responses. Babies therefore have a narrower range over which they can effectively thermoregulate, and this narrow range is shifted to a higher ambient temperature than in adults. In comparison adults can undertake voluntary production (e.g. stamping feet or swinging arms) or acquisition by seeking heat (e.g. being close to a fire, consuming warm foodstuffs, or sunbathing).

Involuntary conservation can be by erecting hairs (by arrector pili) and skin roughening ('goose-bumps' or 'goose-pimples'), decreasing the rate of breathing, or peripheral vasoconstriction (by contraction of pre-capillary arterioles). The primary involuntary means of conserving heat is peripheral vasoconstriction. This enables a progressive centralization of the circulation in order to preserve the core body temperature and hence the efficient functioning of essential systems (e.g. brain, heart). In extreme circumstances the blood flow through the skin can cease, e.g. prolonged cold exposure leads to peripheral vasoconstriction (maximal below 15 °C) interrupted by periodic episodes of locally mediated cold induced dilatation (below 10 °C). This prevents frostbite (necrosis due to tissue freezing), by maintaining an adequate tissue perfusion. Voluntary conservation can be attained by wearing more clothing, curling to reduce body surface area, or avoiding the cold (e.g. sheltering). Heat conservation can also be achieved by bladder voiding, in order to reduce the volume of the body that has to be heated.

Temperature

141

Involuntary removal is by peripheral vasodilatation (by relaxation of pre-capillary arterioles), sweating (by sudoriparous glands), or altering the rate and depth of breathing (by panting). In extreme circumstances the blood flow through the skin can increase ten-fold to 3-4 l.min^{-1}. With panting, more rapid breathing (increasing from 12-14 to 40-140 breaths.min^{-1}) increases respiratory evaporation, while shallower breaths (decreasing from 500-1000 to 100-250 ml.breath^{-1}) ensure that the oxygen intake and hence metabolic rate do not increase. Voluntary removal can be attained by wearing less clothing, spreading the body to increase surface area, or seeking cold (e.g. fanning oneself to increase convection). Heat gain can also be actively avoided (e.g. shading oneself).

The amount of heat produced by the body depends directly on the metabolic rate. Increasing metabolic rate solely in order to produce heat is called regulatory thermogenesis. At low temperatures, more epinephrine is secreted, and the liver therefore delivers more glucose into the blood providing more substrate for respiratory metabolism. In addition, an increase in thyroxine progressively uncouples oxidation from phosphorylation so that the excess glucose is less efficient at producing ATP and releases more heat. Overall, an excess of thyroxine causes an increase in basal metabolic rate. This eventually leads to a decrease in bodyweight, as more calories are used and less fat insulation is needed. Conversely, a deficiency in thyroxine causes a decrease in basal metabolic rate. This eventually leads to an increase body weight, as less calories are used and more fat insulation is needed. If increases in metabolic rate can not meet thermoregulatory demands, then over time fat deposition occurs to increase the thickness of insulation. This increase in body weight is also associated with a lower surface area to volume ratio. For both of these reasons less heat is lost. Hence less heat needs to be produced, and thermoregulation can be maintained at a lower metabolic rate.

With increasing metabolic rate the core temperature rises. Meanwhile, skin temperature falls with the evaporation of sweat. The temperature gradient between the core and skin increases. During exercise the core temperature can be as high as 40 ^{o}C and the skin surface as low as 30 ^{o}C. This increased gradient increases heat loss from the body, by 150-900 W. Any increase in metabolism that is not due to a need to raise body temperature, will produce additional heat that will need to

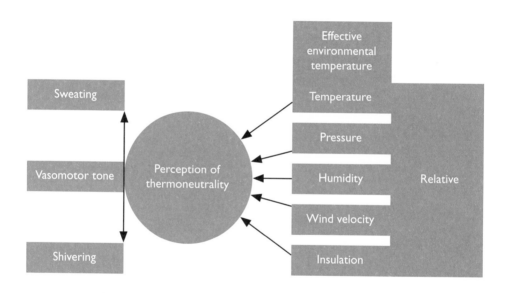

be lost in order to maintain a normal body temperature. Thus maximum metabolic performance is limited by the removal of heat. The removal of heat usually primarily depends on the rate of sweat evaporation and in turn the rate of water replacement. Therefore, the upper temperature of thermoregulation is usually limited by dehydration.

THERMAL COMFORT:

Overall, thermal comfort depends on the relative temperature, humidity, pressure, wind velocity and insulation. All of these factors are interchangeable in their contribution to the effective environmental temperature and the perception of thermoneutrality. In the thermoneutral zone between shivering and sweating, it is changes in vasomotor tone that continue to contribute to the regulation of body temperature. At rest with a relative humidity of 50 %, thermal comfort is at an effective environmental temperature of: 25-26 °C when lightly dressed, 28 °C when naked, and 31-36 °C when in water. (NB the greater variability in water is due to an increased reliance on conduction and hence on differences in insulating fat.) For an adult at rest, outside effective environmental temperatures of 0-50 °C involuntary actions are inadequate and behaviour must change. For example, a 70 kg naked adult at rest can only increase metabolism from 1 W at 28 °C to a maximum of 5 W at 0 °C (cf. 3.5 kg babies can increase from 2 W at 33 °C to 7.5 W at 23 °C). Above an effective environmental temperature of 50 °C, more sweat is secreted than can be evaporated, 100 % of the skin surface is wet, and surplus sweat can only drip from the body. In addition, any behaviour that increases the metabolic rate will thus also decrease the effective environmental temperature for thermal comfort.

CHANGES IN THERMOREGULATORY BALANCE:

Acclimatization to changes in environmental temperature can take the form of tolerance or adaptation. Tolerance involves short-term tactics, while adaptation involves long-term strategies. Short-term responses to changes in temperature occur within seconds or minutes. These involve changes in metabolic rate, vasomotor tone, rates of breathing and sweating or shivering, and tactical behaviour. Long-term responses to temperature change occur over periods longer than a day. These involve changes in basal metabolic rate, fat deposition, hair growth, sweat composition, sweating or shivering threshold, and strategic behaviour.

Temperature

145

Heat tolerance is the short-term acclimatization to high temperature.
The high temperature can either be due to the environment (e.g. the
tropics), or due to extreme endurance exercise (e.g. athletes can
have rectal temperatures of up to 42 °C when running a marathon).
With heat tolerance, the threshold temperature at which sweating
starts is decreased and the rate of sweating is increased. There is then
a corresponding increase in thirst and water intake. In the short-term
this decreases the dependence of thermoregulation on changes in
vasomotor tone and so maximizes cardiovascular efficiency.
However, water loss is increased and predisposes to dehydration.
To maintain body temperature under sustained extreme conditions,
thermoregulatory responses need to be modified further.

Heat adaptation is the long-term acclimatization to high environmental
temperature. Basal metabolic rate is decreased, so that the
body produces less heat, and hence less heat needs to be lost by
thermoregulation. In contrast to heat tolerance, with heat adaptation
the threshold temperature at which sweating starts is increased
and the rate of sweating is decreased. Therefore the body is less
prone to dehydration. In addition, the electrolyte content of sweat is
decreased, and so electrolytes are conserved. There is also an increase
in haemoglobin concentration to compensate for acute episodes of
decrease in plasma volume. The body is therefore also better able to
cope with dehydration if it occurs. In a humid environment the rate
of sweating decreases further still (hidromeiosis). This enables the
conservation of water that would not be able to evaporate if secreted
as sweat.

Cold tolerance is the short-term acclimatization to low environmental
temperature. There are no known systemic modifications of
thermoregulatory responses. However cold induced vasodilatation
occurs, whereby a localized brief restoration of peripheral blood flow
every 7-15 minutes helps to prevent frostbite. Cold adaptation is the
long-term acclimatization to low environmental temperature. With cold
adaptation the basal metabolic rate can be increased by 25-50 %. This
occurs in combination with decreases in the threshold temperatures at
which both shivering and short-term increases in metabolic rate start.

Local changes can also occur. On occasion, the extremities can feel
either cold and clammy or hot and dry. This can be due to differences

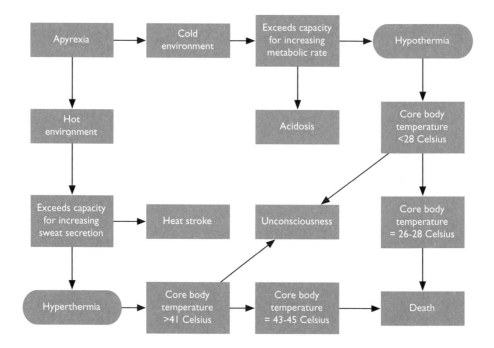

in exposure or activity that cause selective changes in vasomotor tone and sweating. Alternatively, it can sometimes be attributed to selective autonomic stimulation. For example, the 'fight or flight' response would be augmented if the palmar (hand) and plantar (foot) surfaces were wet, as sweaty hands and feet grip better due to increased surface tension. Modification of the thermoregulatory response can also occur by shifts in the optimum body temperature. During infection, cytokines (inflammatory mediators) are released. Some of these chemicals are pyrogenic (e.g. interleukin-1, interleukin-6, prostaglandin E) and lead to disruption of the hypothalamus, causing a rise in the 'set point' of the body 'thermostat'. The body therefore detects itself as being too cold, vessels vasoconstrict and shivering occurs to raise the temperature (i.e. a fever occurs). A high body temperature (pyrexia or febrility) is thus suggestive of infection. On return to normal when the fever 'breaks', the body is too hot, leading to vasodilatation and sweating. Overall, fever can be considered an overreaction of the thermoregulatory response. In the elderly it is possible for the opposite of fever to occur, whereby the 'set point' for body temperature can fall to as low as 35 °C.

HYPOTHERMIA AND HYPERTHERMIA:

Normal body temperature (when the person is said to be apyrexic or afebrile) has its limits, and large deviations cause extreme effects. Abnormally low body temperature is referred to as hypothermia (below 35 °C). The lower limit of thermoregulation is primarily dependent on the maximum rate of metabolism. Prolonged periods of high metabolic rate give rise to both respiratory and metabolic acidosis. With hypothermia shivering occurs when core temperature reaches 34 °C, amnesia at 33 °C, sleepiness at 30 °C, unconsciousness at 28 °C and death from myocardial fibrillation at 26-28 °C. Maximal metabolic rate cannot prevent hypothermia either when naked in air below 0 °C, or when submerged in water below 10 °C. Hypothermia may be induced, for example during transplant surgery, in order to decrease basal metabolic rate and hence oxygen requirement, thus decreasing structural damage when cardiovascular circulation is suspended. Some living tissues survive various degrees of freezing better than others. Cryopreservation can be used to slow metabolic reactions towards a state of near complete suspended animation (most successful when cooling is gradual and re-heating is rapid). This can be used to store viable cells (e.g. sperm), transport organs for transplant, or

For written notes:

149

save severed extremities for re-grafting. If the whole body is cold, cardiovascular and respiratory arrest cannot clinically confirm death: "you're not dead until you're warm and dead".

Abnormally high body temperature is referred to as hyperthermia (above 39 °C). The upper limit of thermoregulation is primarily dependent on the maximum rate of sweat secretion. In the immediate short-term, body temperatures of 42-43 °C can be tolerated. However if hydration is insufficient, such that sweating can not restore normal body temperature, then heat stroke can occur over 90 minutes. With heat stroke, the progressive death of neurons causes a series of problems. The sequence of events with hyperthermia may include some or all of the following: disorientation, delirium, muscle twitching (resembling shivering) and convulsions occurring at around 40 °C; unconsciousness occurs at 41 °C and death at 43-45 °C. This whole sequence of events is self-reinforced to some extent (i.e. there is positive feedback). Excess core heat encourages an increase in metabolic rate, which produces more heat. This is also accentuated by heat damage to the cells that control thermoregulation, the main counterproductive effect of which is to decrease sweating.

SUMMARY OF TEMPERATURE:

The efficiency of hydration and hence volume regulation is affected by deviations in temperature. An increase in temperature will generally result in an increase in either volume or pressure or both, due to increased molecular kinetic energy and expansion. Furthermore, changes in vasomotor tone for temperature regulation compete with the need to maintain cardiovascular pressure for volume regulation. Compensation by thermoregulation has additional implications. For example, every degree that the body temperature rises above normal causes an additional 500 ml.d^{-1} of insensible water loss, and this in turn affects hydration. The solubility of blood gases also decreases with increasing temperature, thus affecting oxygenation. Increases in temperature also affect both the structure and function of proteins by denaturing, thus affecting the energy yielded by respiratory metabolism, and hence causing acidosis. With decreasing temperature, inadequate perfusion due to vasoconstriction can lead to accumulation of lactic acid, and hence cause acidosis. Therefore both increases and decreases in temperature can also upset acid-base balance.

For written notes:

151

CHAPTER 8.

TIMING

INTRODUCTION TO TIMING:

Timing is the regulation of actions in relation to others to produce the best effect. The timing of life in all its aspects is termed chronobiology. In particular, the relationship between supply and demand is one of acute importance for life. It is necessary that active processes are co-ordinated to prevent both surpluses and deficits, either of which leads to inefficiency and compromised competitivity. The timing of one activity with respect to others has obvious implications for survival. Expectation with respect to the temporal structure of the environment can lead to an increase in efficiency. Internal biological clocks ensure optimal timing based on synchronization with environmental cues (such timing cues are often termed Zeitgebers). Periodic variations in autonomous cellular processes are thereby reflected in rhythms of alternating activity. These endogenous rhythms are not a direct mirror of external fluctuations, but continue independently of stimulation, and are capable of advancing or retarding the timing (changing the phase) in order to adapt to changing conditions. Many internal processes are involved, such as the frequency of nerve impulses and the rate of protein synthesis. It is at the metabolic level that these temporal processes are perpetuated, and it is the nervous, endocrine, and reproductive systems that are largely implicated.

PHYSICS OF TIMING:

The unit of time is the second (s). One second in time is equal to the duration of 9,192,631,770 periods of the radiation corresponding to the transition between two hyperfine levels of the ground state of the Caesium-133 atom. One minute is equal to 60 seconds, and one hour is equal to 60 minutes. The mean solar day is no longer used to derive 86,400 seconds of time, because of the continuous deviation in day length from 24 hours.

For written notes:

153

CHEMISTRY OF TIMING IN THE BODY:

With respect to the cellular basis of timing, there are physical
constraints on the time that is required to fulfil certain requirements.
The rates of diffusion, intracellular and intercellular transport will
all have a part to play in limiting timing. Ultimately the origin of any
endogenous rhythm should be traceable back to a cycle of biochemical
activity that is located in cytoplasm and under nuclear control. However,
endogenous rhythms could even be perpetuated by the continuous
repetition of certain physical processes. Metabolic processes, secretion
and cell cycle division (possibly via a negative feedback loop pivoting
on transcription activity) could all be implicated in a 'clock' mechanism.
The cycling of protein accumulation and depletion can be used to
regularly trigger other processes.   For example, one mechanism for
timing intracellular events may be the rate of deamidation of glutaminyl
and asparaginyl residues in peptides and proteins. The underlying rates
of different activities may even work together to determine different
aspects of a single rhythm. Indeed, there may be many different clocks
for the many different rhythms that are demonstrated by the body.
There may also be internal clock centres which regulate particular
rhythms by changing hormonal or neural activity. Such centres with
responsibility for the control of timing processes are the adrenal, pineal
and pituitary glands. One clock centre may have control over more than
one rhythm. Furthermore, one endogenous rhythm may be influenced
by more than one clock centre, and the degree of contribution may vary
depending on the prevailing conditions. It is not yet known whether
such centres act purely as originators, pacemakers, or as only part of a
more extensive mutual synchronization system.

155

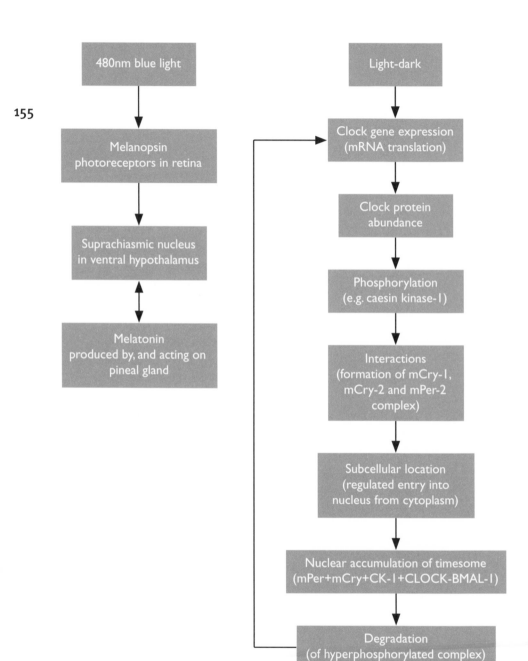

TIMING BY THE BODY:

There are many ways in which organisms can demonstrate rhythmicity. However, they all have one thing in common, ultimately the oscillations are in some way linked to the rotation of the planet with respect to the moon and sun. It is the inherent predictability and environmental dominance of these planetary cycles that results in them being the predominant synchronizing stimuli. The entraining signals can be either directly or indirectly from one (or more) of the daily, tidal, or seasonal cycles. The *periods* of all these cycles vary with geographical latitude as well as the phase of the planetary motions with respect to each other (a period is the interval between successive phase reference points). Thus it is expected that all endogenous rhythms will inherently have a degree of flexibility that enables re-entrainment depending on the prevailing time and place.

The internal clocks appear to be located in the anterior hypothalamus, especially the suprachiasmatic nucleus (for sleep-waking) and the ventromedial nucleus (for temperature and feeding). The pineal gland is indirectly light-sensitive and hence under photoperiodic control. In this case, pinealocyte secretion of the hormone melatonin (N-acetyl-5-methoxytryptamine) is inhibited by light and stimulated by dark. The dominant central pacemaker appears to be the suprachiasmatic nucleus, which receives photic input from photosensitive melanopsin ganglion cells (underneath rods and cones in the retina) via the retinohypothalamic tract. There is a key transcriptional feedback loop in pacemaker cells, whereby rhythmic binding of negative regulators to DNA-anchored CLOCK-BMAL1 heterodimers generates a rhythm in transcriptional activity. The post-translational negative regulators are the period genes (mPer-1 implicated in phase delays, and mPer-2 implicated in phase advances) and cryptochrome genes (mCry-1 and mCry-2). Overall there is a hierarchy of distributed oscillators, which use specific genetic, biochemical, hormonal, neural and behavioural signal cascades in order to elicit systemic physiological rhythms.

Rhythms appear to underlie all biological processes. Diurnal (daytime activity) rhythms such as the sleep-wake cycle, body temperature and blood pressure have one cycle every 24 hours. These tend to lead to a peak in activity at around 3 pm in the afternoon and trough at about

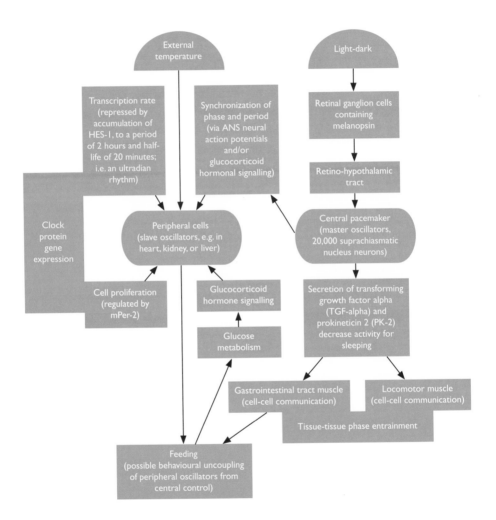

3 am in the morning. Cortisol secretion also has a diurnal rhythm, but with spikes in secretion every 1-3 hours and also further secretion in response to meals and stress. Maximum cortisol blood levels occur before waking and are 450-700 nmol.l^{-1}, while minimum levels occur after falling asleep and are 80-280 nmol.l^{-1}. Increases in cortisol prepare the body for activity, while decreases permit recovery and healing during inactivity. The diurnal rhythm in cortisol level thus overlaps the diurnal rhythms of waking, temperature and blood pressure, but with more frequent cortisol fluctuations superimposed upon that daily cycle. Ultradian rhythms are defined as being shorter than 22 hours (about one day), and include: cortisol secretion, respiration rate, heart rate, electroencephalogram activity, gastro-intestinal secretions and motility, and renal excretion. Infradian rhythms are defined as being longer than 26 hours (about one day), and include: sperm production, ovulation and menstruation (and hibernation and migration for species which undertake these activities).

Take the example of sleeping. The duration of daylight depends on the time of year, therefore there is a need to adapt sleeping patterns. Each morning there is a tendency to wake up slightly earlier than the previous morning, and this time can be fine tuned by resuming sleep if conditions are appropriate. This confers an advantage to survival as 'the early bird catches the worm'. This sleeping pattern is disturbed during 'jet-lag' following transmeridian flight. The day-night stimulus is shortened by a flight to the east, or lengthened by a flight to the west. Diurnal rhythms require about one day for every 1-2 hours of time zone traversed in order to regain their original phase. Re-entrainment takes less time after flights to the west than after flights to the east (i.e. resynchronization is quicker after a delay in phase than after an advance). Also different rhythms can require varying times for adjustment. For example, work and social activity patterns adjust more quickly than temperature and autonomic functions.

In humans, the circadian (about one day) periodicity of the sleep-wake cycle develops between 15-20 weeks after birth. This is entrained by many factors in addition to light, such as social contact and feeding. When integrating these changes with other body processes damping can occur, this reduces the contrast between the different states associated with the rhythm. If on the other hand the body is separated from environmental synchronization and uniform conditions are sustained,

159

THE SPEED OF THINGS:

Actin-myosin sliding	15 mm.s^{-1}, 5 times.s^{-1},
Auditory reaction time	150-250 ms
cf. velocity of sound in air	331.45 m.s^{-1} (at 0 ^{0}C)
Axoplasmic flow	200-400 mm.d^{-1}
Bile secretion	600 ml.d^{-1}
Blood flow	0-30 l.min^{-1}
Blood velocity	0.05-0.7 m.s^{-1}
Cell division (once committed)	11-24 hours per cycle
Cilia beating (each moving 5 µm)	5-10 beats.s^{-1},
Conduction of action potentials	
Myocardial cells	0.05-3.5 m.s^{-1}
Unmyelinated axon	0.5-2 m.s^{-1}
Myelinated axon	2-160 m.s^{-1}
Conscious thought	10-100 bits.s^{-1}
Contraction of striated muscle	0.2-8 ms
Contraction of myocardium	200-300 ms
Diffusion of small organic molecules	
through cytoplasm	50 µm.s^{-1}
Energy consumption by body	6-100 MJ.d.$^{-1}$
Flatulence (gender and age	
independent, less when	
asleep, more after eating)	
10 times a day	500-1500 ml.d^{-1}
Follicle regeneration	
Eyebrow	56-73 d
Scalp	112-139 d
Thigh	77-240 d
Gastric secretion	up to 3 ml.min^{-1}
Gastro-intestinal peristalsis	30-120 cm.min^{-1}
Gastro-intestinal transit time	2.5-5 days for 10 m
Glomerular Filtration Rate	1 ml.sec^{-1}.m^{-2} body surface area
Female	88-128 ml.min^{-1}.1.73m^{-2}
Male	97-137 ml.min^{-1}.1.73m^{-2}

then the period of timing eventually reaches a steady state and the rhythm is said to be free-running. Thus the removal of entraining stimuli such as the onset of darkness can result in changes in period length. These free-running (or un-entrained) endogenous rhythms are usually slightly longer than the daily light-dark cycle. The range for body temperature is 23-27 hours, and for motor activity is 20-33 hours. Sustained deviations from the anticipated pattern of daily activity can create a hormonal imbalance that leads to psychiatric disorders (e.g. insomnia, seasonal affective disorder and depression). There can be further implications for health and safety, which are particularly relevant to shift workers. Nocturnal meals are associated with lipid intolerance and insulin resistance, and these factors in turn increase the risk of heart disease, while sleep disturbance and sleepiness compromise attention.

The direction and degree of any effect on an endogenous rhythm can be dependent on the points that the cycle is both interrupted and resumed, and the duration and magnitude of the intervention. For example, the point in the sleep-wake cycle at which the light stimulus ends will determine the size of a phase shift, whereas continuous light will shorten the time taken for the sleep-wake cycle to fade-out. Fading can be defined as a decreasing difference between the extremes of a cyclical response. Fading may be due to a divergence in the phase of a rhythm between different parts of the body, so that processes become independent of each other. This loss of synchronization has even been demonstrated in different cells of the same organ (e.g. kidney). Thus instead of reinforcing oscillations in activity, there is a dampening due to an accumulation of errors. This flexibility in periodicity allows for a limited adaptation to variations in exogenous stimuli. Ultimately however, it is changes in temperature that are capable of having a universal effect on biological rhythms, as the underlying timing mechanisms all depend on intracellular chemical reactions whose rate will be temperature dependent.

It is essential that an individual body keep a tight constraint on endogenous rhythms. This is supported by the fact that variability in rhythm periodicity within an individual is typically of the order of minutes, while that between individuals can be measured in hours. Within an individual, even a small alteration in the periodicity of a rhythm could have serious consequences for the timing of an event.

Hair growth (combined total)	40 m.d^{-1}
Eyebrow	0.14-0.16 mm.d^{-1}
Scalp	0.31-0.41 mm.d^{-1}
Thigh	0.13-0.25 mm.d^{-1}
Heart rate (at rest)	60-100 beats.min^{-1}
	lower = bradycardia
	higher = tachycardia
	(2-3.10^9 beats in lifetime)
Locomotion (whole body)	up to 9 m.s^{-1}
Membrane transport	
Channels open for	1 ms can convey 100 ions
Carriers take	10 ms for a single two-way exchange
Monosynaptic stretch reflex arc	5-30 ms
Nail growth	0.2 mm.d^{-1}
Oxygen consumption by body	0.2-1 l.min^{-1}
Respiration rate	4-140 breaths.min^{-1}
	12-14 breaths.min^{-1} at rest
Saliva secretion	0.1-7 ml.min^{-1}, 1.2 l.d^{-1}
Single nerve impulse	1 ms
Skin growth	2 months for 60 μm
Sweat secretion	300 ml.d^{-1}, up to 1.6 l.h^{-1}
Synaptic transmission	0.5 ms for 20 nm
Urination (rate decreases with age)	
Female	15-18 ml.s^{-1}
Male	12-20 ml.s^{-1}
Ventilation	6-100 l.min^{-1}
	6-8 l.min^{-1} at rest
Visual reaction time	200-300 ms
cf. velocity of light in vacuum	299.792458 Mm.s^{-1}
Water consumption by body	1.5-10 l.d^{-1}

To complicate things further, many active processes must have the flexibility to integrate, in order to achieve any desired goal for the body as a whole with maximum efficiency.

DECELERATION AND ACCELERATION:

A biological clock can be decelerated or accelerated due to inhibition or induction of the timing process. The degree of change that will cause an adverse consequence depends on the timing process as well as the process being timed. Shifts in time may also affect the ability to keep time. Desired reference 'set point' values may be disrupted and reset differently. This ability to adapt is integral to the timekeeping process. Adaptation requires a mechanism for both down-regulation and up-regulation. Such changes in rate may depend on the: availability of nutrients for utilization, growth of cytoskeletal components (e.g. mitotic spindle), production of proteins by endoplasmic reticulum, production of vesicles by golgi apparatus, or transmission of nerve impulses. If timing is too slow, then supply will be too late to meet demands efficiently. There can be delays because a longer metabolic or neural pathway is followed, cell division is retarded, or secretion is impaired. Conversely, if timing is too fast, then supply will be too early to meet demands efficiently. There can be advances because a shorter metabolic or neural pathway is followed, cell division is promoted, or secretion is enhanced. These processes may be implicated in the timing and co-ordination of themselves, each other, or higher physiological processes.

A delay or advance in a metabolic process can culminate in an adverse physiological effect. For example, causing either a delayed or premature change in vasomotor tone. This could then lead to a detrimental deviation in: growth rate, metabolic rate, respiration rate, heart rate, or rate of renal excretion. However, these physiological rates are determined by an overlapping need to control: volume, hydration, oxygenation, temperature, pH, and ultimately energy. A change in one physiological process can thus have numerous and far reaching effects. Therefore, the complex interconnectivity of the body leads to many incidental consequences, even in response to only a single change in metabolic timing.

For written notes:

163

SUMMARY OF TIMING:

The supply of substrate, rate of production, and demand for product, can all independently be either too slow or too fast. Insufficient supply can lead to a deficit in product, and excess supply can lead to a surplus of product. Low demand can lead to a surplus in substrate, whereas high demand can lead to a deficit in substrate. Similarly, a high rate of production can lead to inadequate availability of substrate and excess accumulation of product, whereas a low rate of production can lead to a surplus of substrate and a deficit of product. Thus there must be a great deal of co-ordination to time the fulfilment of needs for a variety of processes, and there must be at least some limited flexibility in the range of acceptable levels. It appears likely that the needs of the body are co-ordinated by many rhythms, involving the nervous, endocrine, cardiovascular and muscular systems, in order to orchestrate the appropriate communication and motility.

The many rhythms of the body are integrated so as to optimize conditions for the whole body. For example, low body temperature coincides with sleeping at night-time. Deviations from balanced values in any aspect, will initiate drives to fulfil those needs. Frequently, there will be a feedback loop involving hormonal secretion or neural excitation facilitating the promotion or inhibition of an appropriate activity. Importantly, the timing mechanism is inbuilt and not learned. It is known that there is an inherited genetic basis to the internal biological clocks; although a period of fluctuating environmental conditions is often needed for synchronization before the inherited timing mechanism becomes fully operational. Therefore, the absence of an expected endogenous rhythm due to a detrimental intervention can usually be rectified by an appropriate entraining stimulation.

For written notes:

165

CHAPTER 9.

REPLACEMENT

INTRODUCTION TO REPLACEMENT:

Replacement is defined as taking the place of something, and can occur by supersedence, substitution or restoration. Biologically, replacement involves a continuous cycle of conception, growth and death. In theory cells, organisms, populations and species all reach an optimum number in their growth at some stage in time. For all of these, the loss of some is essential for the efficient gain of others. Moreover, the processes of living, do by necessity, result in erosion of the body supporting those processes. There is therefore a continuous balance between *construction* via processes such as cellular division and expansion, and *destruction* via processes such as attrition and failure. Thus, in practice the process of living is dynamic and forever changing. So life can never actually reach equilibrium, as this would signify an end to life. There are many seemingly diverse processes involved in maintaining life by replacement. These include: conception, nutrition, growth and development, drives such as thirst and hunger, sleep, health, wound healing, damage and even death.

CONCEPTION:

Replacement of the whole body is called reproduction and this always requires a point of conception. Conception is the inception of new life. Cellular combination followed by cellular division and growth is *the* mechanism of conception and reproduction. Sexual reproduction requires the interaction of two separate genders, male and female. Genetic material from both genders is combined in order to conceive a new individual. Male and female are complementary so that this interaction can take place. The genetic material is in the form of gametes, males producing sperm (the smallest human cell) and females producing ova (the largest human cell). The male transfers sperm into the female reproductive tract during the process of sexual intercourse (copulation, coition, insemination). Following this the fusion of male and female gametes forms a zygote, the progenitor of the next generation (one generation is equal to half an average lifetime, or about 35 years). The combination of sperm and ova is the culmination

Replacement

For written notes:

167

of the joining of two organisms that have successfully survived to adulthood. Following this process of fertilization (syngamy) there is a new individual, which must then undergo growth to develop into an adult itself. This growth and development commences in the female reproductive tract. The zygote differentiates through various stages and implants (nidates) in the uterine endometrium. The female is then pregnant (gestating) and a placenta is formed for blood exchange (typical of viviparous organisms which give birth to live young). Thus it can be seen that reproduction requires the reconstitution of old living materials within the bodies of the parents. As such, reproduction is strictly speaking not an act of creation.

Cell division normally involves the mitotic 'cloning' of diploid cells. The production of gametes involves the meiotic division of diploid into haploid cells, containing half the normal number of chromosomes. Gametogenesis occurs in gonads, which are the last organs to develop. During puberty sexual dimorphism becomes apparent. The accessory reproductive organs (breasts) develop along with secondary sexual characteristics such as differences in vocal cords, hair and fat distribution. Females produce relatively few ova, each with a large energy store. Oogenesis (ovogenesis) commences before birth producing about two million follicles, although only 400-600 follicles will ovulate during a reproductive lifetime. Males produce comparatively large numbers of spermatozoa, each highly motile (there are normally over 20 million sperm in each ml of ejaculate; about 100 million sperm in the few ml of each ejaculation). Spermatogenesis continues throughout adult life, the production of sperm taking 56-74 days.

Humans are polyoestrous, with more than one menstrual (oestrous) cycle each year. The menstrual cycle takes 21-35 days, and entails the development of a Graafian follicle from an ovary, to produce an ovum that is released during ovulation. With consecutive cycles one egg is normally released down alternate fallopian tubes. The whole process is under hormonal control. Oestrogen stimulates the oestrous ('period of heat'). This is terminated by the secretion of progesterone from the corpus luteum (when not pregnant), or the placenta (when pregnant).

Sperm have limited resources that cannot be replaced and so their lifespan is proportional to their metabolic rate. Therefore sperm are not activated until ejaculation, when prostate (not prostrate) gland

FOR WRITTEN NOTES:

secretions initiate sperm motility. On reaching an ovum, the acrosomal prominence at the apical end of the spermatozoon makes contact. Fertilizin on the surface of the egg reacts with the antifertilizin on the surface of the sperm (cf. antigen-antibody reaction), thus promoting the agglutination of other sperm and preventing fertilization by more than one sperm.

Reproduction permits the perpetuation of the genetic blueprint for an organism. More than this, the purpose of reproduction is to provide a mechanism and opportunity for modification and evolution. Hence both innovation and optimization of the balance between the many integrated processes of life can occur. Favourable attributes can be retained while unfavourable attributes can be removed. This has the aim of maximizing the efficiency of successive generations of offspring. Thus reproduction facilitates the continuation of life through time. We do not live forever because immortality would restrict the introduction of changes necessary for adaptation. Without adaptation a species would loose competitivity and become obsolete.

NUTRITION:

For replacement of any kind to occur there are nutritional requirements. The structure and function of the gastro-intestinal tract plays a key role in the supply of energy and other resources necessary to fuel replacement. After all, usually nutrients must first be fed into the mouth. The time-course of food transit is partially dependent on the distance travelled. Following on from the mouth, the oesophagus is about 0.3 m long with a transit time of about 10 s. Next is the stomach, which is also about 0.3 m, but has a transit time of approximately 3 h. This then leads on to 6.5 m of small intestine, which takes about 7 h to pass through. Lastly, there are some 3 m of large intestine and rectum, which can take between 2 and 5 days. Therefore, there is a retention time of several days, covering about 10 m of tract. This tremendous length of gastro-intestinal tract is responsible for the digestion and absorption of the diet. Efficient digestion and absorption depend on the adequate diversion of blood to the gastro-intestinal tract, and this will have to be tempered against other essential processes (e.g. maintaining blood pressure and thermoregulation).

Elemental Composition of Body as Percentage of Body Weight	
Oxygen	65
Carbon	18
Hydrogen	10
Nitrogen	3
Calcium	1.5
Phosphorous	1
Potassium	0.35
Sulphur	0.25
Sodium	0.15
Chloride	0.15
Magnesium	0.05
Iron	0.0004
Iodine	0.00004

Aside from water; fibre, protein, carbohydrate, fat, minerals and vitamins must all be replaced to meet the demands of body function. The required amounts of the different dietary constituents will vary according to age, weight, gender, exertion and disease. It is also important to bear in mind that the gastro-intestinal tract will not necessarily absorb the entire dietary intake. For these reasons, estimated average requirements (EARs, recommended daily allowances or RDAs) can include a generous safety margin, which can vary according to who is making the recommendation. The daily recommendation for fibre is 21-38 g. This constitutes the indigestible dietary components (such as cellulose), and ensures efficient movement along the gastro-intestinal tract. The daily recommendation for protein is 38-56 g, half of which must be as essential amino acids. The *essential* amino acids (those needed which can not be synthesized by the body) are: leucine, lysine, isoleucine, methionine, phenylalanine, threonine, tryptophan and valine. Protein is nitrogen-rich and provides construction material for growth and repair, as well as being needed to produce enzymes and hormones for secretion. Protein deficit leads to impaired growth, muscular atrophy and oedema. There is a protein reserve of 40 g in muscle and a further 5 g in the liver and blood. The daily recommendation for carbohydrate is 100-130 g, to provide a readily utilizable energy source. Carbohydrate deficit leads to hypoglycaemia, metabolic disturbances, reduced performance and loss of weight. There is a reserve of 300-400 g as glycogen. In extremis, gluconeogenesis enables a carbohydrate requirement to be substituted by twice as much protein. The daily recommendation for fat amounts to about 30-60 g. Two-thirds of this fat intake contributes to 20-35 % of body energy requirements. Excess fat provides for energy storage, and so gluttonous days can be utilized to balance lean days. The remaining third must be as unsaturated essential fatty acids (such as linoleic acid, which is a precursor for arachidonic acid and in turn prostaglandins). This must also include 12 mg as fat-soluble vitamins (which are A, D, E and K). A decrease in dietary fat can thus lead to deficiencies in fat-soluble vitamins. Essential fatty acids are also important for producing the phospholipids needed for membranes. Finally, there is a daily requirement for about 8 g of minerals and 90 mg of water-soluble vitamins (two-thirds as vitamin C). Minerals and particularly vitamins, provide an integral part of metabolic co-factors and are important for wound healing and immunity; while mineral salts also enable the maintenance of pH, ionic and osmotic balance.

173

Vitamin A (retinol or axerophthol, derived from carotenoids such as β-carotene) has a daily recommendation of 0.9 mg, with a typical reserve of 1-2 years. Vitamin A is important for growth, keratinization of epithelial tissues and formation of the visual purple pigment rhodopsin A. Vitamin A deficit causes xerophthalmia, pharyngitis, gastroenteritis, bone abnormality, hypokeratinization, nerve degeneration, corneal vascularization and night-blindness.

Vitamin B1 (thiamin(e) or aneurin) has a daily recommendation of 1.2 mg, with a reserve of 6-10 days. Vitamin B1 is important for carbohydrate metabolism, especially decarboxylation. Deficit causes an accumulation of pyruvic acid leading to beri-beri and polyneuritis.

Vitamin B2 (riboflavin or lactoflavin) has a daily recommendation of 1.3 mg, with a reserve of 1-6 weeks. Vitamin B2 is an important constituent of flavoprotein enzymes and several dehydrogenases. Deficit causes ariboflavinosis, characterised by flaking of the skin.

Vitamin B3 (niacin, niacinamide, nicotinic acid, sometimes called Pellagra Preventing factor or PP) has a daily recommendation of 16 mg, with a reserve of 1-6 weeks. Vitamin B3 is important for carbohydrate oxidation. Deficit causes pellagra, characterised by dermatitis, darkening and thickening of the skin, and paralysis.

Vitamin B4 (choline chloride) has a daily requirement of 550 mg. Vitamin B4 is important for controlling fat and cholesterol, and nerve transmission as it is a component of acetylcholine. The body has a limited capacity to produce choline and so deficit is rare. When it occurs, deficit tends only to aggravate pre-existing problems with fat metabolism and nerve transmission.

Vitamin B5 (pantothenic acid) has a daily recommendation of 5 mg, with a reserve of 6-9 days. Vitamin B5 is a component of acetylcoenzyme A, and is important for lipid and carbohydrate metabolism. Deficit causes paraesthesia (abnormal sensation) and postural hypotension.

Vitamin B6 (pyridoxine group: pyridoxal, pyridoxol and pyridoxamine) has a daily recommendation of 1.3 mg, with a reserve of 5-7 weeks. Vitamin B6 is used to produce erythrocytes and 5-hydroxytryptamine, and is important for amino acid and glycogen metabolism. Deficit causes seizures in infants, inflammation of the tongue and peripheral neuropathy.

Vitamin B7 (biotin, sometimes called vitamin H) has a daily recommendation of 0.03 mg, with a reserve of 1-6 days. Vitamin B7 is important for the breakdown of pyruvic acid to oxaloacetic acid. Deficit causes lethargy, loss of appetite and dermatitis.

The essential mineral elements are: calcium, chloride, cobalt, copper, fluoride, iodine, iron, magnesium, manganese, molybdenum, phosphorus, potassium, sodium, sulphur and zinc. Calcium, phosphorus and to a lesser extent magnesium are important for bone and teeth formation, glycolysis and pH balance. Sodium, potassium and chloride ions are found in body fluids; sodium extracellularly and potassium intracellularly, while chloride is important for gastric digestion. Sodium is also important for pH balance, while potassium is also important for muscle function, and calcium for self-firing neural tissue. Phosphates are an essential component of phospholipids. Iron is an essential component of haemoglobin and cytochrome, and iron deficiency is a common cause of anaemia. Iodine is an essential component of thyroxine and consequently metabolic rate; deficiency causes hypothyrodism and goitre. Zinc is essential for carbonic anhydrase and insulin, and hence carbohydrate metabolism. Cobalt occurs in vitamin B12 and is essential for the production of erythrocytes and insulin. Sulphur occurs in some essential proteins, such as glutathione. Molybdenum is essential for xanthine oxidase and consequently nucleic acid metabolism. The daily dietary intake for an individual adult should include about 4700 mg of potassium, 2300 mg of chloride, 1500 mg of sodium, 1000 mg of calcium (equivalent to about 25 mmol, only 25 % of which is absorbed), 700 mg of phosphorus, 420 mg of magnesium, 11 mg of zinc, 8 mg of iron (18 mg during menstruation), 4 mg of fluoride, 2.3 mg of manganese, 0.9 mg copper and 0.15 mg of iodine.

Vitamins are essential organic co-factors for metabolism. The majority of vitamin reserves are stored in the liver. Under normal conditions, vitamin D is synthesized in the skin by a photochemical reaction involving ultraviolet light from the sun (UVB, 290-310 nm). In the long-term, inadequate exposure of the skin to sunlight predisposes the individual to osteoporosis (although in the short-term, excessive exposure of the skin to sunlight predisposes the individual to malignant melanoma). The vitamins K and H are produced by the gastro-intestinal microflora, and so can be adversely affected by antibiotics. Niacin can be produced from the essential amino acid tryptophan, if tryptophan is available in sufficient amounts. If the intake of any nutrient does not match the needs of the body, such that an excess or deficit causes an adverse effect, then this is malnutrition. Nutritional deficits lead to characteristic diseases. A lack of protein leads to kwashiorkor,

Vitamin B9 (folic acid, folate or folacin, sometimes called vitamin M) has a daily recommendation of 0.4 mg (particularly important for all women planning and during the first 3 months of pregnancy), with a reserve of 2-4 months. Vitamin B9 contains p-aminobenzoic acid and is important for growth. Deficit causes megaloblastic anaemia (megaloblasts are abnormally large erythrocyte precursors; if they fail to mature anaemia develops). Vitamin B9 is considered particularly important for expectant mothers to prevent congenital neural tube defects (e.g. spina bifida) at the end of the fourth week of development following conception.

Vitamin B12 (hydroxocobalamin or cyanocobalamine) has a daily recommendation of 0.0024 mg, with a reserve of 3-5 years. Vitamin B12 is important for erythrocyte and myelin formation. Deficit causes megaloblastic anaemia and funicular myelosis (degeneration of spinal cord white matter).

Vitamin C (ascorbic acid or dehydroascorbic acid) has a daily recommendation of 90 mg, with a reserve of 2-6 weeks. Vitamin C is important for protein metabolism, and in turn wound healing and the maintenance of connective tissue. Deficit causes scurvy. The daily requirement for vitamin C can increase to 1-2 g with severe trauma or burns.

Vitamin D (D2 is ergocalciferol, D3 is cholecalciferol, and D4 is dihydrocalciferol) has a daily recommendation of 0.005 mg (particularly important in the absence of exposure to ultraviolet light). Vitamin D is important for the absorption of calcium and phosphate from the intestine, and is thus antirachitic (prevents rickets). Deficit causes the incomplete calcification of bones (rickets) in children, demineralisation of bones (osteomalacia) leading to a decrease in bone mass and density (osteoporosis) in adults, and dental caries (tooth decay).

Vitamin E (α, β and γ tocopherol) has a daily recommendation of 15 mg. Vitamin E is an antioxidant and thus protects lipid membranes from peroxidation, and is important for placental function and spermatogenesis. Deficit causes sterility.

Vitamin K (K1 is phyllochinone or phylloquinone, and K2 is farnochinone or menaquinone, sometimes called anti-haemorrhagic factor) has a daily recommendation of 0.12 mg. Vitamin K is important for the production of the clotting factors II (prothrombin), VII, IX and X. Deficit causes impaired blood clotting, along with spontaneous haemorrhage and anaemia.

Vitamin P (bioflavonoid or citrin) maintains the resistance of capillary walls to permeation. Deficit causes capillary haemorrhage. No daily recommendation has been established as yet.

characterised by being 60-80 % of expected weight and having oedema. Whereas a lack of carbohydrate leads to marasmus, characterised by being less than 60 % of expected weight and without oedema. Combined protein-energy undernutrition is called marasmic kwashiorkor and is characterised by being less than 60 % of expected weight with oedema. Being 60-80 % of expected weight without oedema is classified as underweight. Acute undernourishment leads to wasting (low weight for height), while chronic undernourishment leads to stunted growth (low height for weight). A vitamin deficit is called hypovitaminosis (cf. avitaminosis refers to complete absence). Lack of a specific vitamin typically leads to a characteristic deficiency disease.

Nutritional excess leads to either excretion or storage of the surplus. Excess storage can pose some severe problems. Vitamin excess is called hypervitaminosis. A daily intake of 4 mg fluoride is necessary for caries prophylaxis, whereas 10 mg or more is toxic and causes osteosclerosis (abnormal increase in bone density). More than 35 mg of vitamin A, 5 g of vitamin C, 35 mg of vitamin D, or 3 g of niacin in one day can cause a variety of epithelial, bone, gastro-intestinal and kidney disorders, along with anaemia, and in extreme cases collapse. More than 10 g of sodium chloride in one day can cause high blood pressure. Excess protein leads to gout and kidney stones, due to an accumulation of uric acid. Excess carbohydrate or fat, leads to elevated blood cholesterol and the accumulation of adipose tissue. This in turn eventually leads to obesity and various conditions associated with inactivity; particularly atherosclerosis, cardiac infarction, stroke, gout and diabetes mellitus. Obesity is associated with a doubling of the relative risk of death from cancer and a four-fold greater relative risk of death from cardiovascular disease. Excess body weight is estimated to shorten the lives of the obese by an average of 9 years. A body mass index of 25 or more is considered overweight, 30 or more is obese. A body weight 20 % greater than ideal is thus also obese, whereas double the ideal body weight is morbidly obese. These body states are the self-inflicted cause of serious diseases that reduce life expectancy, and can be reconciled by eating less (or more specifically absorbing less nutrients) and exercising more (i.e. avoiding gluttony and sloth).

Obesity due to over-eating (hyperphagia, gluttony) is typically associated with either eating for comfort because of depression, or conversely as a reflection of surplus resources and success.

For written notes:

177

On the other hand, undernutrition due to under-eating (hypophagia) causes starvation (extreme hunger) and emaciation (extreme leanness), and is typically associated with deprivation and extreme poverty, or an aspiration towards emaciation as an ideal body image (as with anorexia nervosa and bulimia nervosa). Depending on the environmental conditions and the level of activity it is possible to survive a maximum of four weeks without food (but only four days without water). With undernutrition about 50 % of body weight is lost before thermoregulation finally fails and body temperature falls just before death. The conditions that lead to eating too much, or eating too little, have equally devastating consequences, ultimately leading to an early death. Thus it can be seen that these states reflect a balance between the success and failure of the individual in a wider sense.

GROWTH AND DEVELOPMENT:

Growth can involve an increase in size or increase in numbers, or both. An increase in the number of organisms is called reproduction, and growth is used to describe the processes of development that follow conception. Strictly speaking, growth describes quantitative increase, while development describes qualitative change. Growth chiefly depends on the increase in cell numbers via hypertrophy and an expansion of cell size via hyperplasia. Growth can be measured as a change in mass, volume or surface area; preferably all increase concurrently. Surface area most closely reflects change in size, but it is impractical to measure. Unfortunately fluctuations in metabolism and particularly in water balance, confound the accurate measurement of weight. Height alone is a poor indicator of overall size. However, it is indeed possible to approximately relate these last three parameters, using the formula of Dubois and Dubois:

$$SA = 7.184 \times 10^{-3} \times W^{0.425} \times H^{0.725}$$

Where; SA = surface area in m², W = weight in kg, and H = height in cm. Wallace's rule of nine gives an approximation for dividing the surface area of the body (used for estimating burns). The head and neck together or an arm each represent 9 %, a leg represents 18 % and the trunk represents 36 %.

FOR WRITTEN NOTES:

There are processes other than growth that enable the development of form. Changes in shape can occur due to selective pre-programmed cell death via apoptosis. Apoptosis ensures that DNA is fragmented to reduce the risk of affecting neighbouring cells. It also ensures that cytoplasmic contents are not released as this would cause inflammation and sequestering of the autoimmune response. Apoptosis can occur where a scaffold has been used as a framework for construction and the scaffold subsequently needs removal. For example to remove cells from between the digits and thus form the fingers. Development also encompasses the conversion of one cell type into another cell type via metaplasia. In general, growth is not localized at one location in the body but can originate from cells occurring at any location. There are however some locations deserving special mention for their capacity to produce new cells. For example the epithelia of the skin, gastro-intestinal tract and uterus have germinative layers in order to continually replace cells sloughed from the free surface. Similarly, the testes produce sperm and the bone marrow produces erythrocytes and leukocytes.

180

Cells with a high rate of mitosis fall into two broad categories, those with a long half-life or short half-life. Cells with a long half-life (i.e. years) include liver cells (hepatocytes). Cells with a short half-life (i.e. weeks) include blood cells, and typically follow a committed path of development from stem cell to terminal differentiation. There is a wide range in the half-life of different cell types. At one extreme, intestinal cells are lost and replaced after only a few days. At the other extreme most neurons last for a lifetime (except for the mere 10,000 brain cells which normally die each day; and the 10,000 brain cells created each day by neurogenesis in the subgranular zone of the dentate gyrus within the hippocampus, and in the subventricular zone of the lateral ventricles within the forebrain). All the other cells of the body fall in between these two extremes. For example, bone is removed by osteoclasts and replaced by osteoblasts about every ten years. Similarly, erythrocytes are produced by bone marrow and degraded in the spleen after about 120 days in circulation (about 2 million are replaced each second, each cell having travelled 1.6×10^5 m in its lifetime). Whereas for skin, the surface is continuously flaking away and the whole epidermis is replaced over about two months. The capacity for cell replacement is thought to be finite. For example, human fibroblasts divide 50 times in culture before dying. Furthermore,

For written notes:

181

the rate of cell replacement varies considerably. Skin loss (and hence growth) is about 30 μm.month^{-1} and varies according to wear (more on hands and feet than elsewhere). Hair growth and nail growth are both about 6 mm.month^{-1}. Overall, it is often quoted that one month is typically required for the integration and subsequent loss (i.e. turnover) of most labelled nutrients.

There must be a fine balance between the loss and replacement of cells or else there would be a change in body mass. However more subtly, inadequate control of replacement mechanisms can lead to damage. For example, insufficient blood clotting leads to haemophilia whereas excess leads to thrombosis. In most cases if the balance of cell proliferation is not well regulated then there will be either insufficient replacement or tumour development. A change in the genetic code of a cell (mutation) can lead to a disorganized development of tissue. Such abnormalities are of genetic origin but can also be acquired through damage. Disorganized development in the embryo or fetus (not foetus) is referred to as teratological. At birth any defect is called a congenital malformation, and any structural abnormality that impairs function is thereafter referred to as a disability.

Development is intricately linked to growth. Changes in body form occur throughout life due principally to different parts of the body growing at different rates and at different times. A structure that grows at the same mean rate as the whole body overall is said to exhibit isometric growth. Examples include the abdominal organs and limbs. A rate of growth that is different to the mean is said to exhibit allometric growth. Examples include the head and genitalia. Of course, depending on the stage in development the same structure may behave either isometrically or allometrically. For example, the ears and nose continue to grow even in old age until death. It is also well known that the nails, skin and hair continue to grow after the death of the body as an integrated whole. This growth slows until eventually decomposition takes over.

The rate of growth changes throughout life according to a familiar pattern. Following conception the first division signifies differentiation into an embryo. The grand growth phase when the majority of growth and development occurs is then underway. After about two months the main organs can be discerned and the embryo can be called a fetus, up until birth (parturition; at around 38 weeks after fertilization)

182

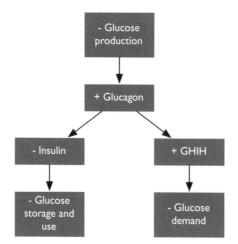

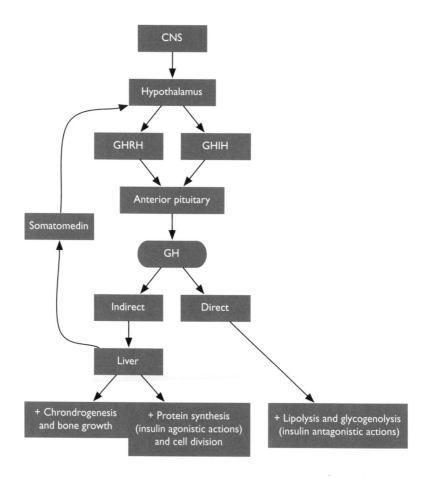

when it becomes a baby (newborn babies up to 4 weeks old are called neonates). With increasing independence the baby becomes an infant at one year after birth. Eventually the infant becomes a juvenile (at around 7 years old). As a teenager (from 13 years old), adolescence coincides with puberty before finally developing into an adult (at 18 years old). Maturation into an adult with the associated fertility and responsibilities signifies the reproductive or constancy growth phase; when there is no further growth, while cell loss and cell replacement are in balance. The maximum limit of body growth is reached following sexual maturity; although localized changes continue to occur, mainly due to varying fat deposition. Maximum skeletal growth is limited by the ossification of the epiphyseal plates at the ends of the long bones.

Growth and the regulation of size are achieved by metabolic processes, and are tempered against ageing by additional defence and recovery processes such as the immune system and sleep. Overall, growth is predominately regulated by the hormones secreted by the hypothalamus, pituitary gland and thyroid gland. The secretion of growth hormone releasing hormone (GHRH) by the hypothalamus increases the secretion of growth hormone by the anterior pituitary. On the other hand, the secretion of growth hormone inhibiting hormone (GHIH, somatostatin, somatotrophin release inhibiting hormone or SIH) by the hypothalamus decreases the secretion of growth hormone by the anterior pituitary. Growth hormone (GH, somatotrophin or STH) directly increases lipolysis and glycogenolysis, increasing blood glucose levels and thereby having an action antagonistic to insulin. In addition, growth hormone causes the liver to produce somatomedins, and thus acts indirectly to increase protein synthesis via cell division in muscle (myogenesis), and bone and cartilage (chondrogenesis). Growth hormone also acts via somatomedins to decrease blood glucose levels, thereby having an acute and indirect action that is agonistic to insulin. Growth hormone is secreted episodically with 3-4 pulses each day, and also in response to fasting hypoglycaemia, certain kinds of stress, vigorous physical exercise and deep sleep. Progression through the various stages of development is controlled by a combination of growth and sex hormones. Meanwhile, the rate of metabolism (and hence growth) is controlled via the secretion of thyroxine from the thyroid gland.

For written notes:

185

Growth is dependent on the efficiency of metabolism, which in turn is predominantly limited by diet. For most cells, the various stages of cell division are typically completed in 11-24 hours, but all are slowed by poor nutrition. In addition under natural conditions there would be a greater demand for energy to be diverted to maintaining body temperature and also to produce fat for insulation, particularly in colder environments. However with time there has been an increasing contribution from lifestyle. When considering choices about lifestyle today, one should take advantage of what our bodies are best designed to do. One should accept that human evolution has been dominated by fulfilling the role of tribal hunter-gatherer (this did not include exposure to uniform environmental conditions, eating three meals a day, or sitting in chairs etc.). Lifestyle interventions that are particularly important today concern exercise (which is usually insufficient, especially when compared with the high number of calories consumed), and advances in healthcare. Consider the fact that on average, African plains people are taller than Western Europeans. One reason is that shorter European women are more likely to survive childbirth because of the intervention of modern obstetrics.

MATHEMATICS OF GROWTH IN NUMBERS:

The mathematical modelling of growth in numbers can be applied to the individual cell, organism, population or species. Growth in any of these cases depends on the balance between generation and degeneration. In reproductive terms, generation depends on fertility and fecundity. Fertility is the ability to produce new individuals, while fecundity is the number of new individuals produced. Similarly, degeneration depends on morbidity and mortality. Morbidity refers to disease, while mortality refers to death. Whether growth is viable and how much growth can be achieved depends on competition for nutrition and space, while also competing against disease and trauma. Excessive competition due to over-population can lead to a population collapse and hence under-population and potential extinction. Growth in numbers is therefore a balance between under-population and over-population.

DRIVES:

The need for the body to replace things gives rise to a number of instinctive responses. These responses are not learned and are

For written notes:

187

base behaviours that ensure that supply will meet demand with respect to the body's fundamental requirements. These are innate homeostatic drives. Indeed there are many such drives, and their action is motivated by both internal and external stimuli. Stimuli reflect the needs of the body, they are detected by the senses and lead to the brain being aware of stimulation (sensation). Classically, the senses are: sight, sound and touch (all physical stimuli); and taste and smell (both chemical stimuli). However, there are also sensory receptors for: balance, pain, proprioception, temperature and visceral (gustatory) function. A change in behaviour motivated by sensory input is a sensori-tactic response. These responses fall into one of two categories, either attraction (associated with pleasure) or avoidance (associated with discomfort or pain). To this end, the brain creates the mind (and the mind is the idea of the body). The mind internally expresses feelings as a balance between pleasure and pain (e.g. happy vs. sad), the external expressions of which are emotions (e.g. love vs. hate). The emotional outcome of this balance depends on whether something is desirable or not, and whether it occurs or not (leading to joy, sorrow, distress or relief). The overriding essence of *being* is to persevere; where striving towards achieving perfection equates to joy. Before this can occur the basic needs of the body must be satisfied, and the motivation to attain these needs is provided by drives. All drives are therefore induced, or inhibited according to need. The most commonly cited examples of such drives are: reproductive (e.g. libido, maternal instincts), thirst and hunger.

Thirst:
Thirst originates from a deficit of about 350 ml in body water (equivalent to 0.5 % of body weight). This also corresponds quite closely to the volume of the stomach at rest or the average cup. Thirst can be either osmotic or hypovolaemic. Osmotic thirst is directly due to a reduction in intracellular fluid volume. The associated increase in intracellular concentration triggers osmoreceptors in the supraoptic nucleus of the hypothalamus. On the other hand, hypovolaemic thirst is directly due to a reduction in extracellular fluid volume. The associated decrease in the activity of stretch receptors in large veins increases the secretion of renin and hence angiotensin II, which in turn acts on the subfornical organ in the forebrain which in turn projects to the paraventricular nucleus in the hypothalamus. The combination of these two types of thirst causes intense thirst. With thirst of any kind

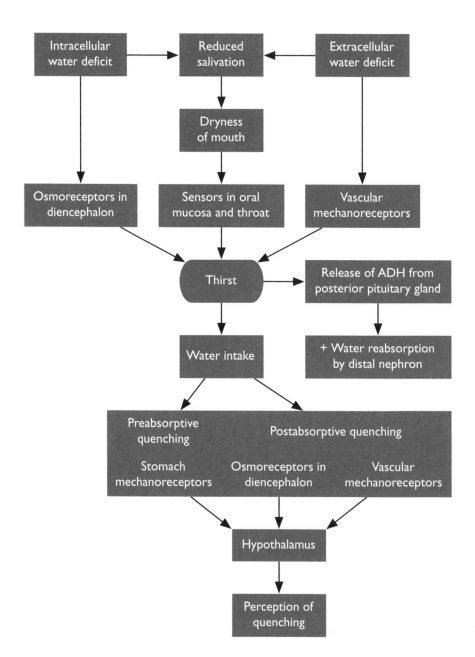

there is a simultaneous secretion of ADH. There is then a reduction in saliva secretion resulting in dryness of the oropharyngeal mucosa. This creates a 'false' thirst, which can also be elicited by physical drying of the mouth and throat. Various clinical conditions can also cause unquenchable thirst through the loss of water faster than it can be replaced. Two such examples are cholera and diabetes insipidus. With cholera, water is lost through vomiting and diarrhoea. Whereas with diabetes insipidus water is lost through large amounts of hypotonic urine.

Hyperosmotic dehydration can occur either due to an absolute water loss (caused by an increase in water loss or decrease in water intake), or a relative water loss (caused by an increase in salt intake or decrease in salt loss). Hypovolaemic dehydration can occur due to haemorrhage or blood donation (isotonic losses). No matter what the cause, thirst is usually quenched by increasing water intake, and this can only be done by either drinking or intravenous infusion. Drinking will stop before there has been time to absorb the water across the gastro-intestinal mucosa and into the bloodstream. This is called preabsorptive quenching and prevents the drinking of too much water and hence overcorrecting the imbalance. Preabsorptive quenching usually closely anticipates water requirements and is followed by postaborptive quenching. There are a number of cues for the sensation of quenching. These may include stretch receptors in the stomach, osmoreceptors in the duodenum, and further stretch receptors in the cardiovascular system. For example, atrial natriuretic factor (ANF, atrial natriuretic peptide, ANP, or atriopeptin) is secreted predominantly by stretching of the right atrium, this increases diuresis (via increased sodium loss), suppresses renin secretion and inhibits thirst. To ensure that other body activities are not disrupted by the need to quench thirst, drinking will also usually occur in advance, in anticipation of the need for water. However, both advance drinking and thirst quenching are dependent on the availability of water. It follows that a reduction in the ability to drink in advance will lead to an increase in thirst.

Hunger:
Hunger originates from a deficit of energy substrates. This refers primarily to the availability of blood glucose, and the buffering action of fat reserves. Thus, force-feeding to gain weight will be followed by under-eating. Gradually, as body weight returns to normal, food intake

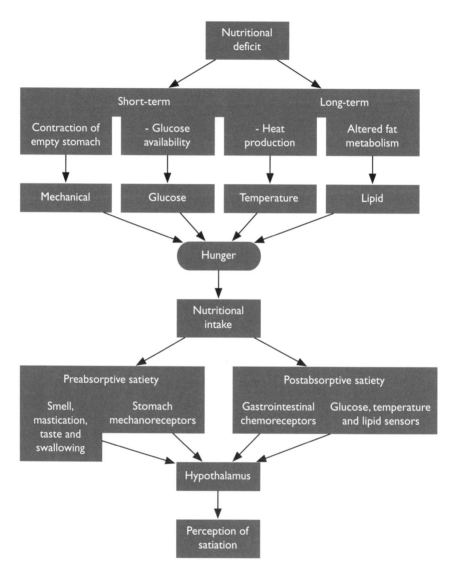

also returns to normal in parallel. With normal body water balance, body weight is determined exclusively by the balance between food consumption and energy expenditure. Hunger and the availability of food are responsible for how much is eaten. Hunger can also be associated with an empty stomach, low body temperature, or low fat reserves. The ultimate consequence of over-eating is obesity, while under-eating leads to emaciation. Obesity occurs due to increased food consumption, decreased energy expenditure, and the deposition of body fat; and is associated with high circulating leptin. Emaciation occurs due to the insufficient intake of energy substrates to meet energy expenditure demands, coupled with the loss of all fat reserves; and is associated with low circulating leptin. Leptin acts as a lipostat; more is released into the bloodstream with increasing body fat, indicating adequate fat reserves. The artificial administration of leptin produces weight loss, in both obese and lean individuals.

Satisfying hunger (satiation) follows a very similar pattern to quenching thirst. Eating will stop before there has been time to absorb the food across the gastro-intestinal mucosa and into the bloodstream. This is called preabsorptive satiety and prevents the eating of too much food and hence overcorrecting the imbalance. Preabsorptive satiety usually closely anticipates food requirements and is followed by postaborptive satiety. There are a number of cues for the sensation of satiation. These may include stretch receptors in the stomach (which expands to 1-4 l when fully distended), chemoreceptors in the duodenum, increased availability of glucose, increased body temperature, and changes in fat metabolism. For example, glucoreceptors and amino acid receptors in the intestinal mucosa, trigger the secretion of cholecystokinin (CCK), which in turn inhibits the action of the stomach in generating hunger. Overall, the hypothalamus functions to balance hunger. The lateral regions are involved in starting eating, while the ventromedial regions are involved in stopping eating. Appetite is the desire to eat certain foods, and this also affects hunger. Appetite is associated with behavioural habits and desires including: frequency of feeding, quantities consumed, cravings (which are often drives to correct nutritional deficiencies), and palatability. These behaviours are not only in response to hunger, but they are also linked to eating in advance, in anticipation of the need for food. However, both advance eating and satisfying hunger are dependent on the availability of food. It follows that a reduction in the ability to eat in advance will lead to an

For written notes:

193

increase in hunger.

Intuitively, we know that there are comfort states for: breathing, excretion, hydration, satiation, sleep, warmth and dryness.
These comfort states are in approximate order of priority, but priorities can overlap depending on the severity of an imbalance. It can be anticipated that this prioritization reflects the balance in emphasis between competing pathways such as oxygenation, hydration, nutrition and thermoregulation. All of these pathways receive feedback from each other, and so the corresponding drives are also inextricably interdependent. There are no drives that directly motivate the maintenance of volume, pH, timing, or energy. However in this context volume is linked to hydration, and pH is a reflection of internal reactive conditions for the generation of energy from nutrition. The timing of intraindividual communication has an overriding influence to ensure that all the needs for replacement are integrated and prioritized. This optimizes energy management within the body in order to maximize the probability of fulfilling the ultimate goal of perpetuating the individual.

SLEEP:

Sleep is a state of reduced responsiveness associated with unconsciousness. More time is spent sleeping than doing any single conscious activity. About 25-50 % of every day is spent asleep, and there is a very strong rhythm associated with the timing of sleeping and waking. The reasons for this use of time remain obscure but some form of recuperation seems obvious. Initially, sleep was probably a means of conserving energy when activity was inefficient. An organism that has evolved to take advantage of daylight will be at a disadvantage in the dark of night-time, and vice versa. In addition, a prolonged suspension of physical activity permits a more uniform and lower metabolic rate. This may provide time for essential and ongoing maintenance (e.g. digestion, growth and repair). Finally with the evolution of higher thought processes, a prolonged suspension of mental concentration on what is physical permits the opportunity for reflection on thoughts and their integration into the consciousness through dreaming.

There are several stages to sleep, all defined by changes in electroencephalography (EEG), but also linked to the threshold for waking. The stages are: awake, transitional, lightest sleep, light sleep,

195

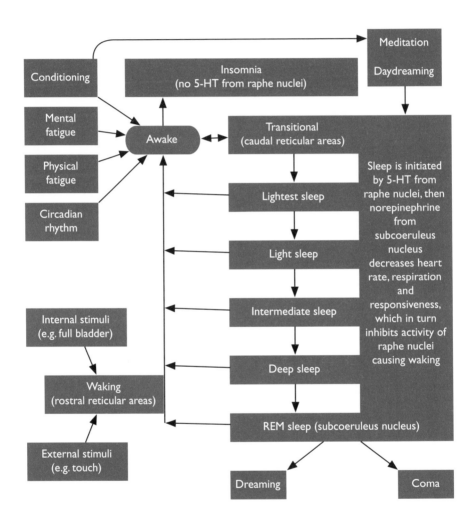

intermediate sleep, deep sleep, and rapid eye movement (REM) sleep. This succession of stages is repeated 3-5 times during one night of sleep. REM sleep is associated with dreaming and is repeated about every 1.5 hours during sleep, with each episode of REM lasting 20 minutes or more. Dreams during the early period of sleep are realistic and related to the events of the preceding consciousness, enabling the consolidation of thoughts. This consolidation then occurs without the distractions of varying external stimuli and the associated varying efforts of conscious attention. Dreams during the later period of sleep tend to be bizarre and unrealistic, enabling speculation on thoughts. In general only the most recent dream is remembered on waking.

Both internal stimuli (e.g. fullness of bladder, headache) and external stimuli (e.g. touch, sound, light, temperature) can affect sleep. However, not all stimuli will cause waking. Stimuli that are below the threshold required to cause waking can be incorporated into dreams. These stimuli are often interpreted in dreams in a way that reflects conscious experience. For example, fullness of bladder may lead to dreaming of urinating into a toilet while in actual fact the dreamer is bed-wetting. In this case there are also sensations of warmth and wetness spreading around the groin and legs that conflict with the dream. Conflicting sensations often result in both confusion and arousal from sleep. Alternatively, dreams alone can cause waking without any other stimuli. Nightmares in particular can be so psychologically disturbing that the individual is woken. With increasing age the total time spent sleeping decreases. Babies can sleep for 16 hours a day, with 50 % of this as REM sleep. By adolescence, the amount of time spent sleeping has fallen to about 9 hours a night, with 25 % as REM sleep. In old age, sleeping typically declines to 7 hours or less a night. However the elderly also take daytime naps more frequently. This is associated with a decrease in stamina and exertion, both physically and mentally.

The depth of sleep affects some autonomic functions, for example causing large variations in heart rate (40-70 beats.min^{-1}), respiration rate (4-14 breaths.min^{-1}) and sexual arousal (erection can occur during REM sleep). Whereas, other autonomic functions such as body temperature do not vary greatly with the depth of sleep. Lighter depths of sleep are needed periodically in order to enable movement and thus

For written notes:

197

promote adequate circulation (e.g. pressure sores start to form after about 45 minutes of immobility, especially at weight bearing bony prominences). Many activities can occur while unconscious during sleep. For example: snoring, talking (somniloquy), grinding of teeth (bruxism), bed-wetting (enuresis), sleepwalking (somnambulism), and the stopping of breathing (apnoea). Paralysis is also possible; where effectively the conscious mind awakes, but the body remains asleep.

Several factors are thought to contribute to the initiation of sleep and waking. Either physical or mental fatigue can increase the drive to sleep; although usually only the rapidity of onset and not the total duration of sleep is affected. The initiation of sleep is also reinforced by conditioned patterns of activity (e.g. brushing teeth, voiding bladder, undressing, turning-off light, head touching pillow). However, the internal initiation of sleeping and waking is predominantly associated with the ascending reticular activating system. Large fluctuations in the activity of the reticular formation in the medulla oblongata both stimulate awakening and maintain consciousness. Indeed, damage to the reticular area can result in coma. Sleep is also associated with a decrease in responsiveness to external stimulation. In this respect, the reticular formation is linked through collaterals from sensory pathways as they pass through the brainstem. Thus small fluctuations in reticular activity are thought to be responsible for changes in the degree of conscious attention and hence daydreaming. Within the reticular formation the caudal areas appear to have an inhibitory effect on the rostral areas. It then follows that the caudal reticular areas initiate sleep, while the rostral reticular areas initiate and maintain waking. Cells towards the medial line of the medulla oblongata in the brainstem are also implicated in sleeping and waking. Here, damage to the raphe nuclei causes insomnia lasting for several days, while damage to the subcoeruleus nucleus causes loss of REM sleep (but not non-REM sleep). In this case, an accumulation of 5-hydroxytryptamine from the raphe nuclei is thought to initiate sleep, while an accumulation of nor-epinephrine from the subcoeruleus is thought to cause the progression through the various stages of sleep. Following from this, the activity of the subcoeruleus inhibits the activity of the raphe nuclei, permitting waking. There is a range of physical factors that may also contribute to the disruption of sleep. These include: light, temperature, sound, touch, fullness of bladder, thirst and hunger. Too much sleep is associated with a lack of physical activity and sloth.

For written notes:

199

In the long-term, excessive sleep leads to sluggish bowel movement and constipation, muscle wastage through under-use, and eventually bone degradation at the major joints. A chronic inability to sleep (insomnia) is associated with anxiety. Individuals who report insomnia frequently lack an accurate knowledge of their sleeping patterns. Many supposed insomniacs sleep more than they think they do and need less sleep than they realize. It is clear from sleep deprivation experiments that 5.5 hours of continuous sleep every night can be sustained without a perceptible impairment of performance. However, if night-time sleep is insufficient then there will be daytime fatigue and sleepiness, which is at best inefficient and at worst dangerous.

HEALTH:

Health can be considered to be the absence of damage.
When maintaining health it is first necessary to decide what level is being considered. Health can be considered at the sub-individual level of cells and the immune system, at the individual level of personal health, or at the supra-individual level of population health.

The immune system provides defence against microbial infection, such as that from bacteria, viruses, fungi and parasites. There are interwoven *unspecified* and *specified* immunity mechanisms. Also involved are repair and regeneration (healing) processes. Pathogen-*unspecific* (inborn) mechanisms rely on phagocytes and lysosomes, which employ cellular mobility and soluble factors respectively, in order to 'clean up' debris. Pathogen-*specific* (adaptive) mechanisms rely on antibodies (there are thought to be between 10^6 and 10^9 different types) that act on the specific antigens of infective agents. Immunological memory is the encoding of information by antibodies for antigens. This provides the basis for distinguishing primary immune responses (having no memory) from secondary immune responses (having memory). The key to the success of the antibody-antigen mechanism is the recognition of 'foreign' vs. 'self' (as 'learned' before birth). Extreme cases result in either immunological tolerance or autoimmune disease. With immunological tolerance, the targets either develop resistance or decrease the immune response. Whereas with autoimmune disease, antibodies are produced that attack proteins of the body.

FOR WRITTEN NOTES:

201

One example of pathogen-specific immunology in action is (perhaps surprisingly) that of blood groups. Erythrocytes and serum from different blood groups have different antigenic properties. Blood group A has antigen A on red blood cells and antibody anti-B in the serum. Blood group B has antigen B on red blood cells and antibody anti-A in the serum. Blood group O has neither A nor B antigens on red blood cells, but has both anti-A & anti-B antibodies in the serum. Whereas blood group AB has both A & B antigens, but neither anti-A nor anti-B antibodies in the serum. Mis-matches between different blood groups in this ABO system lead to agglutination and haemolysis. Those with type O blood can give blood to all other types, in addition all blood types can of course receive their own blood type, and AB can receive any type. Therefore there is a need for cross-tests to check compatibility before transfusions. In addition the states of Rhesus positive and negative occur; due to production of antibodies after previous sensitization to the opposite Rhesus type (via either the placenta or a blood transfusion).

An allergy is a disturbance in the regulation of the immune system which is manifested as an overreaction of the immune response. For example, lymphocytes can become sensitized to a harmless antigen (e.g. pollen). These lymphocytes then release an array of chemicals that act on: blood vessels, mucus glands and sensory nerve endings; causing vasodilatation, oedema and itching. Allergies may be *anaphylactic* (potentially rapidly fatal) such as with insect stings, nut consumption or penicillin; or *delayed* such as with nickel, chrome or poison ivy.

Vaccination can employ one of two types of immunization. *Active* immunization relies on administering antigens before an illness. Whereas, *passive* immunization relies on administering antibodies during an illness. There are some difficult viruses that only give rise to a slow infection, such as hepatitis or herpes. However there are more dangerous viruses that produce no effective immune response, for example Kuru ('laughing death' of Papua New Guinea), Creutzfeldt-Jakob Disease (CJD; where the central nervous system degenerates), and Human Immune deficiency Virus (HIV) which causes Acquired Immune Deficiency Syndrome (AIDS; where lymphocytes are decimated).

For written notes:

203

Personal health is referring to the behavioural habits and choices of individuals that maintain health. This includes hygiene, diet, and avoidance strategies. For example: food hygiene (especially against salmonella), dental hygiene, washing body, washing clothes and cleaning of the living environment (such as bedding, especially to remove parasites). These behaviours can vary in severity from ignorance and neglect, to that required for prevention and prophylaxis, to fear and extreme phobias.

Epidemiology describes disease patterns, identifies contributing factors, and provides management data to control the spread of disease within populations. Population health is dependent on public hygiene maintained by the national health and environmental health services. Environmental health regulates the safe supply of water and food, and the removal of all unhygienic types of waste (sanitation). Complementing this is primary care which is pivotal in preventing the spread of disease. Communicable diseases (contagion) can be epidemic (meaning 'upon the people') and endemic (meaning 'in the people'). The same disease can be either epidemic or endemic for two different populations at the same time, or both epidemic and endemic for the same population at different times. Diseases associated with populations include: amoebic dysentery (amoebic intestinal infection causing diarrhoea), bubonic plague (plague killed 10,000 a day at its peak in AD540, in an outbreak that lasted 50 years; plague also killed 24 million from 1347-52, and remained endemic in Europe for the next three centuries), cholera (bacterial intestinal infection causing vomiting and diarrhoea), CJD, herpes (viral infection of membranes), HIV, legionnaires' disease (bacterial lung infection; named after an outbreak at the American Legion convention in 1976 at Philadelphia), malaria (meaning 'bad air'; sporozoan infection), syphilis (spirochete infection; first described in 600 BC in China), tuberculosis (bacterial infection), typhoid fever (Salmonella typhi bacterial infection causing intestinal inflammation) and typhus (rash and fever due to rickettsias bacterial infection; transmitted by the bites of lice, ticks and fleas). When considering disease in populations, generally no more than 90 % of the population die, while the remainder survive as resistant 'mutants'. If the whole population were decimated then the disease agent would effectively self-destruct. In other words, by killing all potential hosts a disease removes its own means of propagation. In this way, duels between infective and infected

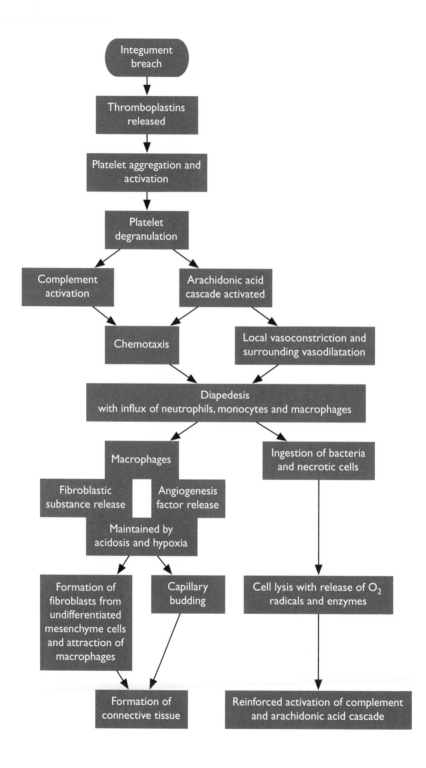

organisms reach a balance.

In the diagnosis of disease, recognising abnormality requires recognition of normality. Common problems occur commonly, so exclude these first. In other words, consider possibilities in order of probability. Only once the problem has been correctly identified can appropriate treatment be initiated. When considering the pharmacological treatment of disease it is interesting to note that until the twentieth century there were only a few widely accepted effective therapeutic drugs. These were opium (from the seeds of the opium poppy; a source of morphine) to relieve pain, mercury for the treatment of syphilis, lithium salts for dementia, digitalis (from the seeds and leaves of the foxglove) for heart failure, and quinine (from the bark of the cinchona tree) for malaria. In addition, there was a vast array of lesser-known herbal remedies, some of which were highly effective. Such remedies included seaweed (a source of iodine) for hypothyroid goitre, and willow bark (a source of salicylate) as an anti-inflammatory agent.

WOUND HEALING:

A wound is a physical injury, typified by a breach in the integument. A wound can be due to a contusion (crushing and bruising), laceration (tearing) or incision (cut). The capacity for replacement depends not just on the type and extent of damage, but also on the complexity of the structure that is damaged. Cells can be categorized as static, labile or stable. Static cells have very little or no adult proliferative capacity (e.g. cardiac muscle, neurons). Labile cells continuously replace those lost (e.g. exfoliated epithelial, degraded haematopoietic). Whereas, stable cells have a low mitotic rate until they are lost (e.g. smooth and skeletal muscle, connective and glandular cells). If a damaged tissue cannot be restored then it is typically substituted with connective tissue. The replacement process of wound healing has several phases. The phases of wound healing are: inflammation, organization, proliferation, epithelialization and contraction.

The series of events that cascade immediately after injury are collectively termed inflammation. Any tissue undergoing inflammation is generally indicated by the suffix 'itis', for example dermatitis refers to inflammation of the skin. All inflammatory responses encompass: vascular, tissue and chemical events. The vascular events follow

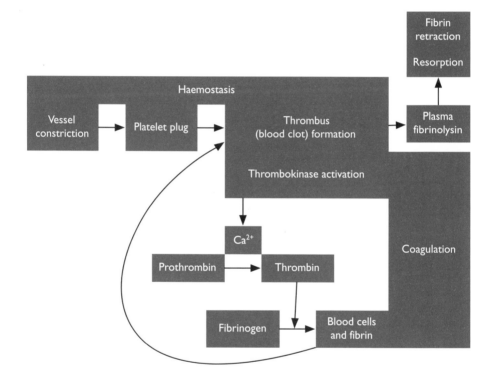

the order: arteriolar dilatation, increased haemokinesis, increased capillary permeability, haemoconcentration and finally haemostasis (where vasoconstriction and platelet degranulation predominate). The tissue events follow the order: transudation (exudation of fluid from blood vessels), pavementing (margination of neutrophils between membranes), diapedesis (leukocytes moving out of venules), haemorrhage and thrombosis. The chemical events include the release of various vasoactive mediators such as: histamine, 5-hydroxytryptamine (5-HT or serotonin), bradykinin, neutrophils, and various polypeptides. There are many different types of inflammation, but all share the same five cardinal signs. Those being: *calor* (heat), *rubor* (redness), *tumor* (swelling), *dolor* (pain; strictly speaking this is a symptom) and *functio laesa* (loss of function). It is interesting to reflect that menstruation can be considered a form of physiological inflammation.

Prior to actual cell replacement a wound will undergo organization (demolition). In this context, organization is the replacement of solid inanimate material with granulation tissue. Replacement is stimulated by the release of cellular constituents during the breakdown (autolysis) of dead cells. Cell breakdown also provides nutrients for local recycling in the construction of new cells. Necrosis destroys the connective tissue framework that has no regenerative capacity (as in myocardial infarction). Alternatively, inflammatory components fail to resolve (as with severe pneumonia or large thrombi). In most cases, there is a period of several days during which debris is removed from the wound and cells such as fibroblasts migrate to the wound.

Wound healing next depends on cell proliferation. There are six main types of cell proliferation, four reversible and two irreversible. The reversible forms of proliferation are hypertrophy, hyperplasia, metaplasia (transmutation or substitutive growth) and dysplasia (disorganized growth). The irreversible forms of proliferation are neoplasia (excessive new growth) and anaplasia (growth of indistinguishable cell types) which are persistent, progressive and purposeless. Benign and malignant tumours are both irreversible forms of growth. The distinction between them is that malignant cells are 'aggressive' and will migrate, spreading to distant sites (metastasizing); whereas benign cells will not. Usually, neoplastic growth is benign, and anaplastic growth is malignant. However, wound healing is predominantly by hyperplasia. Replacement (in this context regrowth)

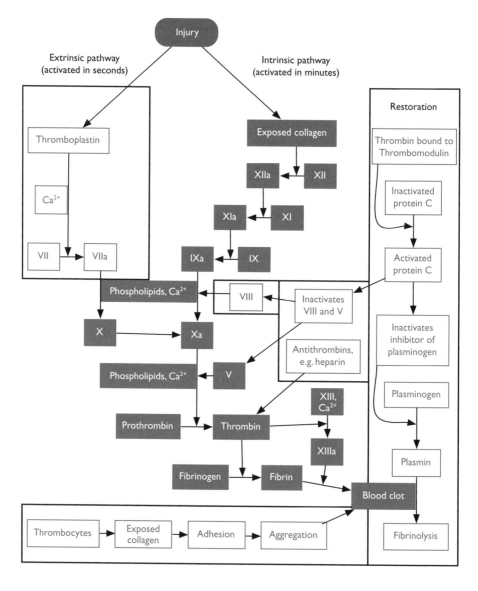

occurs by growth of adjacent cells and is regulated by various hormones.
There is either hyperplasia of the original cell type, or hyperplasia of
migrating fibroblasts. If restoration is not possible, then fibroblasts give
rise to connective tissue as a substitution.

Epithelialization is the covering of the breach by epithelial cells.
These epithelial cells divide and grow from the wound edge. A thin
layer will have covered most wounds within 24-48 hours. This provides
an important protective barrier against further damage and allows
some restoration of function. There is progressive collagenization and
vascularization over a period of weeks. However, adnexal structures
(adjacent or appending; i.e. hair follicles, arrector pili muscles,
sudoriparous and sebaceous glands) do not regenerate, so healing
eventually leads to a pale epithelium covered scar.

Contraction determines the scar dimensions. A circular wound
leads to linear scar, and a rectangular wound leads to stellate scar.
Myofibroblasts are responsible for wound shrinkage. Prolonged wounds
or mechanical stress leads to excessive formation of myofibroblasts and
excessive contraction. A contracture is the limitation of joint movement
due to excessive wound contraction.

The healing of a wound in skin occurs by either primary or secondary
union (intention). Let us consider primary union of a surgical incision.
Exuded fibrin forms a scab at the surface. Neutrophils migrate to the
wound over 3-4 hours, before being replaced by macrophages, which
fill the wound by 24-36 hours. By 7 days an acantholic (hyperplastic
or thickened) epidermis has formed. Then granulation tissue forms with
anastomoses (joining of vessels), myofibroblasts and neovascularization
(angiogenesis); fibroplasia occurs in parallel. In contrast, secondary
union is the healing of an irregular wound. Factors adversely affecting
healing can be either local or systemic. Local factors include inadequate
blood supply (e.g. due to thrombi), infection (e.g. of bacteria after
burns), and movement (e.g. of bones at joints). Systemic factors include
nutrition (e.g. protein and vitamin C deficiencies restrict collagen
synthesis) and iatrogenic toxins (drug side-effects; e.g. corticosteroids,
cytotoxins). In addition, there are several possible complications
associated with healing. These are dehiscence, keloid, cicatrization and
neoplasia. Dehiscence is busting due to sudden mechanical stress
(e.g. coughing after an abdominal operation). A keloid is a 'tumour-like'

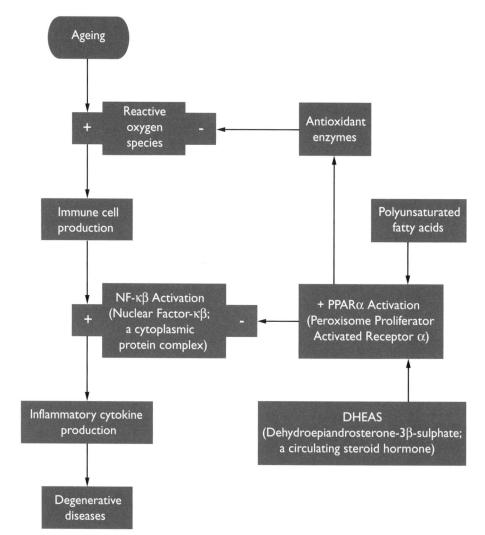

scar due to hypertrophic growth. Cicatrization is an exaggerated contraction causing a deformity (e.g. excessive stenosis; narrowing of vessel or duct). Finally, neoplasia can occur (e.g. 'suture-line' cancer).

DAMAGE AND DEATH:

Damage is caused by the disturbance of normal function to the extent that it exceeds adaptive capacity. There are three principal adaptive strategies that can be employed to compensate for increases in demand. These are, increasing cellular activity (e.g. P450 enzyme induction), recruitment of dormant structural units (e.g. nephrons), or a proliferation of cells (e.g. hepatocytes). When these processes cannot compensate, then pathological damage (disease) results. Damage can be offset by replacement and recovery, which encompasses: repair, regeneration and restoration. Repair is the synthesis and maturation of scar tissue as a substitution for original tissue. Regeneration is the supersedence of necrotic cells by viable cells of the same type. Whereas, restoration is the restitution of the original structure and function.

In pathological terms, cells may be damaged by degeneration or necrosis. Cell degeneration occurs due to slight irreversible damage, whereas cell necrosis occurs due to severe irreversible damage. It sounds obvious, but for death to occur there must first have been life. The recognition of death relies on this. Necrosis is the morphological change that follows cell death, by which cell death is recognised. The one exception is death by myocardial infarction, which may be too rapid for necrosis to occur. Death is a failure of energy expenditure to actively maintain cellular integrity. This can be brought about by a wide variety of factors such as radiation (solar, nuclear), chemicals (toxins), infection (bacterial, viral, fungal, protozoan), impaired nutrition, inadequate blood supply, genetic programming, or trauma. Ultimately, the breakdown of cell membranes and the subsequent dispersion of cellular contents result in a return to equilibrium with the environment.

As the body 'grows old', the net balance of cell replacement becomes negative. That is, cell division decreases and catabolism outweighs anabolism. This indicates progression into senescence, when there is also a decline in physiological efficiency. The loss in fitness of youth is compensated for by the gain in experience of age, although this is only effective up to a point. In addition, with advancing age,

For written notes:

213

a high proportion of body cells have undergone an ever-higher number of replications. With repeated cell division there is an increasing probability of accumulating errors in the DNA. These mutations will then themselves be replicated and may cause abnormal growth. It must therefore be an inescapable truth that the rate of replacement will ultimately limit lifespan. Thus with ageing there is an inescapable rising risk of developing tumours. Not surprisingly the study of geriatrics is closely associated with oncology.

The desire to preserve function with age begs the question: "to use or not to use?". Take the example of an osteoarthritic hip joint. Will continued use wear it out quicker, or keep it working? On the other hand, will restricted use wear it out slower, or stop it working? This is not easy to answer. More use will strengthen supporting muscles, maintain both the range of joint movement and confidence in mobility, but increase the risk of a damaging fall and wear away more bone. Less use will decrease the risk of a damaging fall and wear away less bone, but weaken supporting muscles, reduce both the range of joint movement and confidence in mobility. Obviously, there must be a balance between using and not using. Generally, if things are used carefully they will last for longer, but with a gradually decreasing range of action. However once an activity is stopped, the potential for future use is quickly lost.

Interestingly, life can be prolonged by dietary restriction. Eating as desired leads to an otherwise premature death. In experimental animals survival has been optimized by consuming 20 % less than the amount required for satiation. This increases longevity and also results in a more active life. The extended lifespan is thought to be accounted for by a reduction in the occurrence of degenerative spontaneous lesions, possibly mediated by a hormonal mechanism. So the best prognosis is achieved by not always eating when you are hungry. Eat only four-fifths of your *ad libitum* calorific intake.

Regardless of such efforts the ultimate stage in life will always be death. The signs of ageing become ever more prominent: wrinkles, sagging, dry skin, uneven skin texture, uneven skin tone, loss of colour, slowness, weakness, and incomplete recovery from damage. The combined decline in efficiency of all body functions results in a progressively decreasing capacity to adapt and overcome changes, until a point is

For written notes:

reached at which life is no longer viable. Eventually the tolerance limit of one organ is exceeded, which causes systemic stress and vulnerability to total collapse, especially if other organs are already compromised. Failure of one organ for more than three days is associated with 50 % mortality, two organs with 75 % and three organs with 100 % mortality. Death due to natural causes is therefore characterised by multiple organ failure brought about by extreme old age. A fair lifespan was considered to be three score years and ten (70 years), but improvements in healthcare continue to increase expectations of a longer and healthy old age. Nevertheless, the only certainty of life is eventual death.

SUMMARY OF REPLACEMENT:

Replacement encompasses reproduction, growth, healing and other health maintenance processes in the cycle of life and death. This requires a shifting balance of priorities in order for success overall. There is a background of regulating volume, hydration, oxygenation, pH, temperature and their careful timing in order to manipulate energy. The ultimate objective for this energy is reproduction, but to get there, energy must first be diverted towards the requirements of day-to-day living, growth, maintaining health, and - when necessary - healing. With respect to these uses of energy, motivating drives enable the optimization of supply and demand. All of this needs to be integrated and applied, and for that, our attention must turn to communication.

For written notes:

217

CHAPTER 10.

COMMUNICATION

INTRODUCTION TO COMMUNICATION:

Communication is an exchange of information. This is necessary so
that work can be driven by demands. All active processes need to
interact and thus be integrated. Communication within the body is
itself integrated via a 'central overseer', which co-ordinates information
from disparate processes in order to meet the requirements of body.
The nervous, endocrine, cardiovascular and musculoskeletal systems
are all implicated in this control.

Information needs to be conveyed and things need to be moved around,
these two related processes both depend on interaction. Interaction
is fundamentally important for all biological systems. Interactions can
occur within the body and between the body and the environment.
The central nervous and musculoskeletal systems facilitate responses to
external stimuli. The autonomic nervous, endocrine and cardiovascular
systems facilitate internal communication and transport between cells,
tissues and organs. There is a clear distinction between communication
and transport. Communication involves a stimulus-response. This can
be either slow as with hormones, or fast as with nerves. Transport is
not signal bearing, and can bring (nutrition) or take away (elimination).
Broadly speaking there are general and specific mechanisms for both
communication and transport. The endocrine system is an example
of general communication, and the nervous system is an example of
specific communication. While the cardiovascular system is an example
of general transport, and the musculoskeletal system an example of
specific transport. These processes will be considered in more detail in
this chapter in order to gain an appreciation of how the processes in the
preceding chapters integrate to meet the needs of the body as a whole.

All the distances involved are huge considering the relative sizes
involved. Nevertheless the time taken varies depending on the
distribution networks that have evolved. The communication links are
either fast signals or slow messengers. Fast signals include: cAMP, Na^+
(via pumps), Ca^{2+} (via channels), membrane potentials and synaptic
transmitters (e.g. acetylcholine). Slow messengers include: histamine,

FOR WRITTEN NOTES:

219

prostaglandins, antigens, and growth factors. The time taken depends on the means used, which in turn reflects the importance of the process and the urgency attached to its regulation. Accordingly, nerves react in less than a second, blood takes about a minute to circulate around the body and return to the heart, and hormones may act over days to years.

PROCESSES EMPLOYED:

Diffusion and Osmosis:

The cellular transmembrane transport of water and solutes rely principally on osmosis and diffusion (although vesicle production and phagocytosis also play important roles). Osmosis involves the movement of solvent, whereas diffusion involves the movement of solute particles. Consider the factors governing osmosis and diffusion in turn.

Osmotic pressure is defined as the excess hydrostatic pressure that must be applied to a solution in order to make the diffusion potential of a solvent in the solution equal to that of the pure solvent at the same temperature. Van't Hoff's law of osmotic pressure defines the change in osmotic pressure ($\Delta\pi$), in Pa:

$$\Delta\pi = \sigma . R . T . \Delta C_{osm}$$

Where; σ = reflection coefficient (σ = 0-1; e.g. σ = 0.68 for urea in the proximal renal tubule; o represents completely free permeability, whereas 1 represents complete impermeability to the solute in question; thus isotonicity can only occur when σ = 1), R = 8.3145 J.K^{-1}.osm^{-1} (the molar gas constant), T = absolute temperature (K), and C_{osm} = osmolality of solute (osmol.kg^{-1} H_2O, or osmolarity as osmol.m^{-3} solvent water). Hence, osmotic pressure is proportional to both molecular concentration and temperature.

Diffusion (passive transport) is the movement of an entity in a random direction with a random frequency of changing direction. Thus the distance travelled is proportional to the square root of time (if it takes 1 second to travel 1 μm, then it will take 4 seconds to travel 2 μm, and 100 seconds to travel 10 μm etc.). However, this random movement is typically measured from a high to a low concentration. Here the amount of solute diffusing per unit cross-sectional area is directly proportional to the concentration gradient across that section. Fick's first law of

FOR WRITTEN NOTES:

221

diffusion defines the quantity diffusing per unit time, in $mol.s^{-1}$:

$$dQ / dt = (D.A) / l \times \Delta C$$

Where; D = diffusion coefficient (in $mol.m^2.s^{-1}$), A = area of exchange, l = distance, and ΔC = concentration difference. Factors directly influencing the diffusion coefficient include molecular weight and size, fat solubility and ionic charge. The larger D, A, and ΔC, and the smaller l; the greater quantity of substance diffuses with time. NB The quantity diffusing can be difficult to measure as the concentration gradient changes as diffusion proceeds.

The passage of solutes in and out of cells cannot always be explained by diffusion alone. Depending on the type of cell and the prevailing conditions, the permeability of the cell membrane to different solutes can be either accelerated or decelerated by a variety of processes. In particular, lipophilicity and hydrophilicity limit solubility, while protein binding and ionic charge limit filtration at the cellular tight junctions. These processes affect ease of movement (leading to either accumulation or expulsion) in order to deliberately create an ionic imbalance. The existing ionic balance (particularly of sodium, potassium and calcium) together with the cellular metabolic rate will both affect these processes. These are all active processes and as such they require an energy supply and are saturable. The Michaelis-Menten equation can be used to determine transport rate (velocity, v) of a saturable transport process, in $mmol.s^{-1}$:

$$v = V_{max} \times [S] / (K_m + [S])$$

Where; [S] = concentration of substance transported (in mmol), and K_m (Michaelis kinetic constant) = concentration at half saturation (in mmol).

The specificity and sensitivity of communication by such processes all depends on evolutionary requirements. A small range of distinct targets requires a high degree of specificity, and if only a small quantity of signal is needed to elicit a response then there is a high degree of sensitivity. Of uppermost importance to these mechanisms are the roles of membranes and media; and the changes in equilibrium that occur between two media across a membrane.

FOR WRITTEN NOTES:

223

Electrical Potential:
The ionic equilibrium and resting membrane potential determine the electrochemical potential. The Nernst equation can be used to calculate the equilibrium potential (E_x) of an ion 'x', in mV:

$$E_x = R \cdot T \cdot (F \cdot z_x)^{-1} \cdot \ln ([x]_o / [x]_i)$$

Where; Faraday constant, $F = 9.65 \times 10^4$ A.s.mol^{-1}; z_x = number of charges on ion; ln = natural logarithm, [x] = molar concentration of ion, i = inner and o = outer surface of cell membrane.

It is possible to further derive, $E_x = -61 \cdot z_x^{-1} \cdot \log ([x]_i / [x]_o)$.

For example, $z_x = +1$ for K^+, $+2$ for Ca^{2+}, and -1 for Cl^-. If T = 310 K in the body, $[K^+]_i = 150$ mmol.kg^{-1} H_2O, and $[K^+]_o = 5$ mmol.kg^{-1} H_2O. Then $E_x = -90$ mV.

The electrochemical gradient across a membrane depends on both the difference in membrane potential and the difference in chemical concentration. Electrochemical neutrality requires each cation to be balanced by an anion. Sodium and potassium account for 90 % of all cations, the remainder being mostly calcium and magnesium. Similarly, chloride and bicarbonate account for 80 % of all anions, the remainder being mostly phosphates. When considering electricity and cells, resting membrane potentials in all living cells are typically between 50-100 mV. The cell interior is normally more negative due to an uneven distribution of ions. The cell membrane typically separates two aqueous solutions that are almost equal in electrical conductivity. Usually both solutions have an equal number of ions, but their type and proportions are different. Sodium and chloride comprise about 90 % of external ions and 10 % of internal ions. The balance is made up of potassium ions and a variety of large impermeable negative organic particles. At the resting potential the ions are found in the following internal/external ratios: sodium 1/8, chloride 1/30 and potassium 30/1. Under normal conditions there is a steady movement of sodium ions into the cell that have to be actively 'pumped' back out. NB The imbalance in the distribution of the three ions is restricted to the immediate vicinity of the membrane, and does not extend throughout the entire interior of the cell.

225

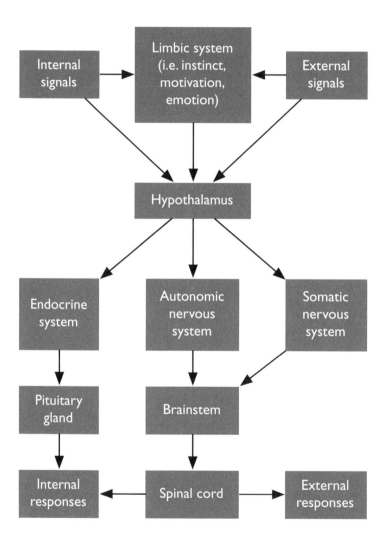

GENERAL COMMUNICATION - HORMONAL:

Hormones are chemical messengers and can be glycoproteins, peptides, phenols or steroids. Hormones are secreted by glands. Secretion through a duct is exocrine (exosecretory; e.g. mammary and sweat glands). If the gland is ductless then the secretion is endocrine (endosecretory; e.g. pituitary and thyroid glands). In addition, if the secretion and target are close together then it is paracrine (affecting adjacent cells), or autocrine (affecting the secretory cell itself). Endocrine hormones are secreted into the bloodstream and so they in particular provide a means for systemic communication. These hormones are carried in blood for the slow and chronic transmission of signals mediated via the circulatory system. Such hormones act on distant target cells in organs, and so require receptor binding sites of high affinity, as typical concentrations are 10^{-8}-10^{-12} mol.l^{-1} (cf. affinity is the tendency to bind, whereas efficacy is the ability to initiate effects). Hormones regulate nutrition, metabolism, growth, reproduction, and also the internal environment. Generally speaking the effect of a particular hormone on a particular target cell is well known, but the mechanism by which that effect is achieved is unknown. Hormones are under the control of a number of different glands that are distributed throughout the body. These are primarily the hypothalamus, pituitary, adrenal, thyroid, parathyroid, thymus and pineal glands. Glands can undergo compensatory hypertrophy or compensatory atrophy, according to sustained changes in demand for the production and release of hormones. Certain other tissues such as the pancreas, duodenum, kidneys and sex organs also have glandular effects.

The hypothalamus exerts control over the other endocrine glands and is especially crucial for the regulation of metabolism. The hypothalamus is a neuroendocrine gland with intimate links to both internal sensory stimuli and behaviour. The hypothalamus is therefore the principal centre responsible for co-ordinating actions that maintain homeostasis. It has a neurosecretory function whereby certain neurons release hormones directly into the blood. The responses of the hypothalamus are predominately orchestrated through the pituitary.

The pituitary (hypophyseal or hypophysial) gland (body) can be divided into four parts; the pars nervosa, pars intermedia, pars anterior and pars tuberalis (NB the whole pituitary gland is the hypophysis).

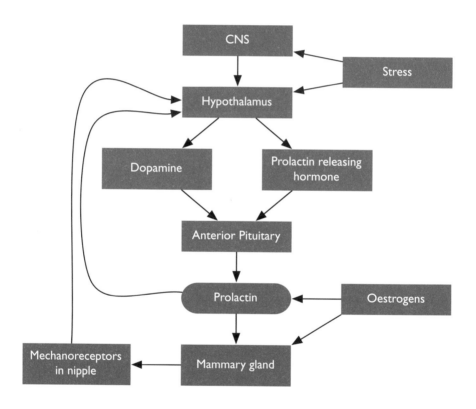

The pars nervosa secretes ADH and oxytocin. These two neurohypophyseal hormones are secreted by the posterior hypothalamus before travelling in an axoplasmic flow along the axons of neurons to the posterior pituitary. The pars intermedia secretes three types of melanocyte-stimulating hormone (intermedin or melanotrophin) called α-MSH, β-MSH and γ-MSH; these increase the production of melanin within melanocytes to darken skin and hair pigmentation (and when released by the hypothalamus these hormones also increase satiety and sexual arousal). The pars anterior secretes growth hormone (GH, somatotrophin or STH), thyroid-stimulating hormone (TSH or thyrotrophin), adrenocorticotrophic hormone (ACTH or corticotrophin), follicle-stimulating hormone (FSH), luteinizing hormone (LH; cf. LH release is triggered by luteinizing hormone releasing hormone LHRH) and prolactin (luteotrophic hormone or LTH). These six adenohypophyseal hormones are all stimulated by releasing factors delivered from the anterior hypothalamus in the blood of the hypothalamic-hypophyseal (hypothalamo-hypophysial) portal system. The first three hormones are associated with growth and metabolism, while the latter three hormones are associated with gender and reproduction. Growth hormone leads to an increase in the rate of protein synthesis. Deficiency in the immature is the cause of pituitary dwarfism. Excess is the cause of pituitary gigantism. This presents clinically as delayed ossification of the epiphyseal plates and hence continued growth of the long bones. Excess in the adult is the cause of acromegaly. This presents clinically as further growth only of acral (extremity) bones such as those found in the skull, jaw, hands and feet. An additional effect of growth hormone secretion is to increase the secretion of glucagon from α-cells in the islets of Langerhans. Hence growth hormone is also implicated in the regulation of blood glucose. Thyroid-stimulating hormone stimulates the thyroid gland to secrete thyroxine. Adrenocorticotrophic hormone stimulates the cortex of the adrenal glands to secrete various corticoids. The three remaining hormones can be described as gonadotrophic. Follicle-stimulating hormone stimulates the development of Graafian follicles in females, or spermatozoa in males. Luteinizing hormone stimulates ovulation and corpora lutea (singular is corpus luteum) formation (luteinization) in the female. In the male, the equivalent hormone is attributed with stimulating testosterone secretion by the interstitial (Leydig) cells of the testis. In females, prolactin stimulates the secretion of milk (lactation) by the mammary glands (oxytocin causes contraction

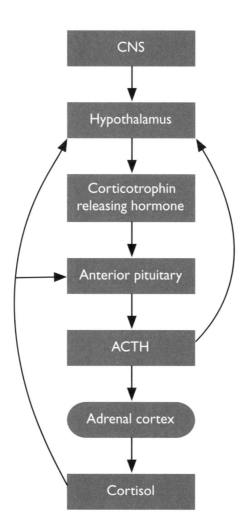

of the myo-epithelial linings of the mammary alveoli, releasing milk within 10-15 seconds of suckling). Prolactin, growth hormone and adrenocorticoids together stimulate the development of the mammary glands. The remaining part of the pituitary gland, the pars tuberalis is not known to have any hormonal secretions.

The adrenal glands (suprarenal glands) can be divided into an inner medullar and outer cortex. The medulla secretes epinephrine and nor-epinephrine (the catecholamines previously known as adrenaline and nor-adrenaline) in response to fright (causing fight or flight), or low temperature. The cortex has three regions and three corresponding types of corticosteroids. Firstly, the 'outer' zona glomerulosa secretes mineralocorticoids such as aldosterone and deoxycorticosterone; in response to renin secretion from the kidney. These are responsible for osmoregulation (along with parathormone from the parathyroid glands and ADH from the hypothalamus). Secretion increases the retention of sodium, chloride and bicarbonate ions, while increasing the loss of potassium and phosphate ions by the kidney. The mineralocorticoids are also implicated in promoting the replacement of damaged tissue. Secondly, the 'middle' zona fasciculata secretes glucocortico(stero)ids such as cortisol (known as hydrocortisone when produced synthetically). These are responsible for the conversion of fat and protein into carbohydrate. Thus resulting in more glycogen stored in the liver, and more glucose released into the blood. The glucocorticoids are also implicated in inhibiting the replacement of damaged tissue (when it would be inefficient to do so during activity). Thirdly, the 'inner' zona reticularis secretes androgens and oestrogens. These are responsible for the development of secondary sexual characteristics that are a balance of male and female respectively. A loss in adrenal gland function is the cause of Addison's disease. As might be expected, this presents clinically as a reduced ability to withstand infection, injury, fatigue and low temperature.

The thyroid gland secretes thyroxine. Thyroxine controls the basal metabolic rate. Insufficient thyroxine is called hypothyrodism and leads to a low metabolic rate, whereas excess thyroxine is called hyperthyroidism and leads to a high metabolic rate. Hypothyroidism in the immature can cause cretinism. This presents clinically as retarded mental and physical development. Hypothyroidism in adults can result in myxoedema. This presents clinically as slow metabolism, atrophy

Hormone	INSULIN	GLUCAGON	EPINEPHRINE	CORTISOL
FUNCTION	Satiation	Hunger	Emergency	Restoration
Glucose Uptake	+	?	+	-
Glycolysis	+	-	+	-
Gluconeogenesis	-	+	+	+
Glycogen	Synthesis	Lysis	Lysis	Synthesis
Fat	Synthesis	Lysis	Lysis	Lysis

of the thyroid gland, lethargy, and swelling of subcutaneous tissues. Hyperthyroidism can manifest as a goitre. This presents clinically as fast metabolism, hypertrophy of the thyroid gland, hyperactivity, and wasting of subcutaneous tissues (an exophthalmic goitre is characterised by protruding eyeballs). Simple goitre is usually due to a deficiency of iodine in the diet. In conjuction with the hypothalamus, the thyroid gland is also implicated in thermoregulation. Low temperature causes thyroxine secretion, which increases metabolic rate, and in excess uncouples oxidation from phosphorylation so that more glucose is oxidised with the release of heat rather than ATP synthesis.

The parathyroid glands secrete parathormone. Parathormone controls calcium metabolism, and consequently influences osmoregulation (in conjunction with ADH from the hypothalamus and mineralocorticoids from the adrenal glands) and ossification. Low levels of calcium ions in the blood promote parathormone secretion. This leads to increased intestinal absorption of calcium, removal of calcium phosphate from bone, retention of calcium by renal tubules and increased urinary excretion of phosphate. Simplistically, insufficient parathormone results in hypoparathyroidism which can lead to nervous irritability and muscular spasm due to hypocalcaemia. Excess parathormone results in hyperparathyroidism, which can lead to osteitis fibrosa. This presents clinically as demineralization of the skeleton, leading to fragile fibrous bones and stones of calcium phosphate in the kidneys.

The thymus gland secretes thymosin which promotes the production of lymphocytes from the thymus gland itself. The thymus is particularly important for the development of the immune response. With age the gland degenerates and synthesis of lymphocytes moves to the lymphatic glands.

The pineal gland secretes melatonin in response to nervous stimulation from the retina. Darkness promotes secretion while light inhibits secretion. Thus the pineal gland is implicated in the control of circadian rhythmicity, specifically the sleep-wake cycle.

The pancreas contains the secretory islets of Langerhans. Each islet contains both α-cells and β-cells. α-cells are small, contain alcohol-insoluble granules, and secrete glucagon in response to low blood sugar, thus promoting the breakdown of glycogen in the liver

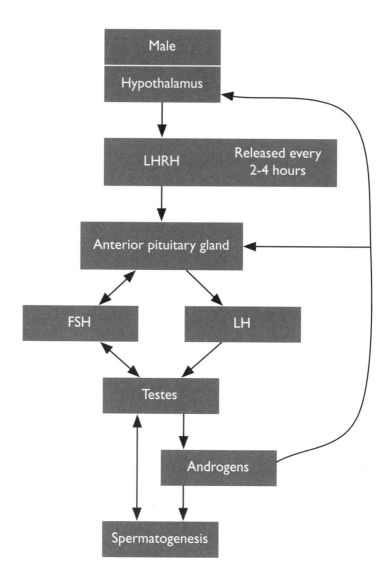

into glucose. β-cells are large, contain alcohol-soluble granules, and secrete insulin in response to high blood sugar, thus promoting the storage of glycogen. Insufficient insulin results in diabetes mellitus. This presents clinically as sugar and ketones excreted in the urine and weight loss. The underlying reason being insufficient glycogen storage leading to the utilization of protein and fat for energy production. The pancreas also functions as an exocrine gland, whereby digestive secretions pass through the pancreatic duct, into the bile duct, before being ejected into the duodenum. The duodenal mucosa itself secretes secretin, cholecystokinin (CCK, formerly pancreozymin) and enterogastrone. Secretin stimulates an increased flow of salt solution from the pancreas (predominately bicarbonate, which acts to neutralize the acidity of bile). Cholecystokinin stimulates an increased flow of enzymes from the pancreas, and also stimulates contraction of the gallbladder and increased bile flow into the duodenum. Enterogastrone inhibits gastric acid secretion and peristaltic stomach contractions, and so decreases the expulsion of chyme into the duodenum. The gastric mucosa secretes gastric secretin, gastrin and enterocrinin. Gastric secretin has the same effect as duodenal secretin. Gastrin stimulates gastric secretions; while enterocrinin stimulates duodenal secretions.

The kidneys contain the secretory juxta-glomerular complex cells. These cells secrete renin in response to low blood sodium ion concentration or low blood flow. Renin acts on angiotensinogen (hypertensinogen) in plasma, breaking it into angiotensin (hypertensin). Angiotensin also acts as a hormone, not only stimulating an increase in vasoconstriction which increases blood pressure, but also stimulating an increase in the secretion of mineralocorticoids such as aldosterone by the adrenal cortex. Aldosterone increases the reabsorption of sodium ions from the distal tubules, which in turn increases the reabsorption of water from the collecting ducts, thereby causing a decrease in urine output and an increase in blood pressure.

In males, the testes contain secretory interstitial cells between the seminiferous tubules. The pars anterior of the pituitary secretes follicle-stimulating hormone (here called interstitial cell stimulating hormone or ICSH) for the maturation of spermatozoa, and secretes an equivalent to luteinizing hormone which stimulates the interstitial cells of the testes. These interstitial cells secrete two androgens called androsterone and testosterone. Both stimulate spermatogenesis, seminal vesicle and

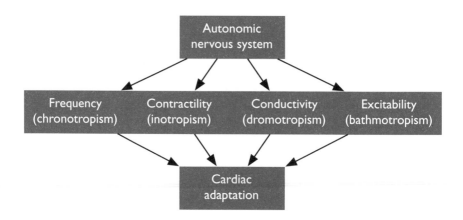

prostate gland activity, secondary sexual characteristics and muscle growth (having an anabolic effect and increasing protein synthesis).

In females, the ovaries secrete the oestrogens, progesterone and relaxin. Again the pars anterior of the pituitary secretes follicle-stimulating hormone, in this instance for the development of Graafian follicles in the follicular phase. The follicular cells secrete a number of oestrogens such as oestradiol (estradiol). These stimulate the preparation of the body for pregnancy. The pars anterior also secretes luteinizing hormone for the repair of the ruptured follicle and formation of the corpus luteum in the luteal phase. The pars anterior also secretes prolactin, which stimulates the secretion of progesterone from the luteal cells, which in turn inhibits further ovulation and promotes mammary gland development. Oestrogens, progesterone and chorionic gonadotrophin are also secreted by the developing placenta and implicated in the continued development of both uterus and mammary glands. The corpus luteum is also implicated in the secretion of relaxin. Relaxin inhibits uterine contraction, but during parturition it stimulates dilation of the cervix and relaxation of the pubic symphysis (while oxytocin causes contraction of uterine smooth muscle).

In addition to all of the above, most tissues have been shown to contain histamine and acetylcholine. Both are released in response to damage and have similar effects. They stimulate the contraction of smooth muscle and also trigger vasodilatation, leading to localized increases in blood flow and capillary permeability.

GENERAL TRANSPORTATION - HAEMATOLOGICAL:

The blood has many other functions aside from the transmission of hormones for communication. The blood also transports heat (by convection), oxygen, carbon dioxide, nutrients and metabolites, has a role in buffering for acid-base balance, and in defence as part of the immune response. The study of these processes and communication between them, therefore requires a thorough understanding of haemodynamics and cardiac adaptation.

There are four ways in which cardiac activity can adapt, all controlled through the action of the autonomic nervous system. The frequency of impulses and thus heart rate (chronotropism), the contractility and thus

EFFECT OF FITNESS ON CARDIOVASCULAR FUNCTION:

At rest, people with different levels of fitness will have similar circulatory requirements and hence have similar cardiac outputs. However, an athletic heart has a lower heart rate at rest, because exercise leads to the development of a larger stroke volume. Thus a fit heart needs fewer beats than an unfit heart to deliver the same cardiac output. In addition, an athletic vasculature has a lower blood pressure, this is because exercise maintains effective vasomotor tone and hence total peripheral resistance is kept lower. Thus a fit vasculature has a lower pressure than an unfit vasculature when receiving the same cardiac output.

force of the heart (inotropism, not iontropism), the conductivity and thus speed of excitation (dromotropism), and the excitability and thus threshold of excitation (bathmotropism). An example combining the cardiovascular system, autonomic nervous system and musculoskeletal system is postural hypotension (a fall in blood pressure with altered orientation). Standing to vertical in cold air, from horizontal in hot water, causes dizziness and fainting. This is due to a decrease in the blood supply to the brain as a consequence of peripheral vasodilatation and gravity, in conjunction with a failure of cardiac activity to adapt quickly enough.

The cardiovascular system consists of a double pump (heart) and cyclical network of pipes (blood vessels). These diverge and converge resulting in corresponding decreases and then increases in diameter. The system contains blood oxygenated by the lungs. The cardiovascular system comprises: arteries, arterioles, capillaries, veins, arterial and venous blood, right side of the heart with the pulmonary circulation, and left side of the heart with the systemic circulation.

Cardiac work (N.m) is the multiple of pressure ($N.m^{-2}$) and volume (m^3). The work of the two ventricles contracting is $1.1 \, J.s^{-1}$ at rest. The pulse wave and distension of vascular walls contribute an extra 20 % (0.22 J). Overall cardiac power is thus about 1.5 W at rest.

The cardiac output is the multiple of heart rate and stroke volume. At rest total cardiac output for a 70 kg human is approximately 70 beats a minute, each of 70 ml (equal to $4.9 \, l.min^{-1}$):

CO = HR x SV

Alternatively, CO = (MABP - CVP) / TPR

Where; CO = cardiac output, HR = heart rate, SV = stroke volume, MABP = mean arterial blood pressure, CVP = central venous pressure, and TPR = total peripheral resistance.

In extreme circumstances the heart rate can increase to give a cardiac output of up to $30 \, l.min^{-1}$. As there is about 5 l of blood in the body, a resting cardiac output of $4.9 \, l.min^{-1}$ equates to all of the blood in the body passing through the heart once every minute. The proportion of

HAEMATOLOGICAL STATISTICS:

Blood accounts for 6-8 % adult body weight.

Systemically this amounts to 4.4-5.5 l,
33-54 % of which is haematocrit (packed red cell volume).

There is typically a red blood cell count of $3.9\text{-}6.5.10^{12}.l^{-1}$,
and a white blood cell count of $4\text{-}11.10^{9}.l^{-1}$.

The white blood cell (leukocyte) count is composed of:
> 67 % granulocytes,
> 27 % lymphocytes, and
> 6 % monocytes.

The approximate composition of 1 mm³ of blood is:
> $4.5\text{-}5.1 \times 10^{6}$ erythrocytes,
> $4\text{-}10 \times 10^{3}$ leukocytes, and
> $0.15\text{-}0.3 \times 10^{6}$ thrombocytes (platelets).

In blood vessels the velocity of flow is inversely proportional to cross-sectional area, typically $0.05\text{-}0.7$ m.s^{-1} ($0\text{-}10$ m.s^{-1} at extremes). Thus the velocity of blood, through an individual vessel, can change with vasoconstriction or vasodilatation and divergence or convergence. However the total volume of flow (m³.s^{-1}), when combined for all vessels, is equal for successive sections. This is defined by Ohm's law, which can be used to derive blood pressure (V), in kPa (where 1 kPa = 7.5 mmHg):

$$V = I . R$$

Where; I = volume flow rate (e.g. cardiac output = 5 l.min^{-1}),
R = resistance (e.g. total peripheral resistance = 2.4 kPa.min.l^{-1}).

cardiac output to different organs is as follows: 24 % to liver and gastro-intestinal tract, 21 % to skeletal muscle, 19 % to kidney, 13 % to brain, 4 % to coronary arteries (supplying the cardiac muscle), up to 9 % to skin (depending on thermoregulatory requirements) and 10 % for the remainder of the body. This division of cardiac output varies according to the predominant activity at the time. For example during heavy work two-thirds of cardiac output is diverted to skeletal muscle, whereas during digestion of a large meal two-thirds of cardiac output is diverted to the gastro-intestinal tract. Thus, blood flow varies with tissue type: skeletal muscle requires $2.0\text{-}4.0 \times 10^{-2}$ ml.g^{-1}.min^{-1} ($0.5\text{-}1.3$ ml.g^{-1}.min^{-1} during exercise), heart requires $0.8\text{-}0.9$ ml.g^{-1}.min^{-1} (4.0 ml.g^{-1}.min^{-1} during exercise), brain requires $0.4\text{-}0.6$ ml.g^{-1}.min^{-1}, liver requires about 1.0 ml.g^{-1}.min^{-1}, and kidney requires about 4.0 ml.g^{-1}.min^{-1}. Compare the approximate proportion of oxygen demand: skeletal muscle accounts for 27 %, liver and gastro-intestinal tract for 23 %, brain for 21 %, coronaries for 11 % and kidney for 7 % of oxygen utilization. The kidneys have the highest blood flow for their weight, but the lowest oxygen demand, because their role is primarily to filter the blood. The distribution of total blood volume (5 l) is approximately as follows: 7 % in the heart, 7 % in the lungs, 7 % in arteries, 7 % in arterioles, 7 % in capillaries, 32.5 % in venules and 32.5 % in veins. The venous vessels therefore act as a reservoir, containing 65 % of the blood in the body. The distribution of vascular resistance is as follows: 19 % due to arteries, 47 % due to arterioles, 27 % due to capillaries and 7 % due to veins.

The Hagen-Poiseuille law defines resistance (R), in Pa.m^{-3}.s:

$$R = 8 . l . \eta . \pi^{-1} . r^{-4}$$

Where; l = length (in m), and η = viscosity (3.5-5.4; in Pa.s or N.m^{-2}.s).

Reducing radius by 16 % doubles resistance. The viscosity of plasma is approximately equal to that of water, however when blood cells are included viscosity is 4 fold higher than water. An important feature of blood flow is that it can be considered to be laminar and as such the viscosity increases with proximity to the vessel wall. Viscosity can change independently of cardiac output. For example, anaemia and polycythaemia are both associated with increased cardiac output, but anaemia decreases viscosity while polycythaemia increases viscosity.

FOR WRITTEN NOTES:

241

The Laplace-Frank law defines wall tension (T), in $N.m^{-2}$:

$$T = P_t . r / \mu$$

Where; P_t (transmural or stretching pressure) = pressure in vessel minus surrounding pressure, r = vessel radius, and μ = vessel wall thickness.

The aorta has both a high transmural pressure and radius. If atherosclerosis weakens the vessel wall, a bulge (aneurysm) forms due to the reduced surrounding pressure. A vicious cycle then ensues: as increasing radius leads to a decreasing wall thickness that leads to an increasing radius, until eventually the vessel ruptures. In contrast, capillaries are 3,000 times smaller than the aorta, and so capillaries have thicker walls relative to their lumen diameter and thus sufficient wall tension to resist pressure.

The regulation of blood pressure plays a key role in maintaining a viable and dependable circulatory system for communication and transport. There are three main factors to consider when measuring blood pressure: magnitude, site and fluctuations. The magnitude of blood pressure varies between the veins and arteries. Venous pressure is low at 0-10 mmHg with small and slow changes (low frequency). Arterial pressure is high at 80-120 mmHg with large and fast changes (high frequency). Blood pressure is the pressure of blood at the height of the heart, relative to atmospheric pressure. Thus there will be fluctuations in blood pressure depending on changes in atmospheric pressure and posture. To compare measurements from different parts of the body it is necessary to correct for the height above or below the heart (by 8 mmHg for every 10 cm of vertical difference). It is also necessary to be aware of the systolic and diastolic rhythm of peaks and troughs in blood pressure due to the beating of the heart. Furthermore there are differences due to whether the blood has been ejected from the right or left side of the heart. The mean pressure in the aorta is 120 / 80 mmHg (= 16 / 10.7 kPa = systolic / diastolic = peak / trough), whereas the mean pressure in the pulmonary artery (artery pulmonalis) is 25 / 10 mmHg (= 3.3 / 1.3 kPa). Blood pressure can be measured either internally or externally. Furthermore these measurements can be either direct or indirect. For example, direct internal measurements can be made by miniature catheter transducers. Whereas, direct external measurements can be made by transducer via a hypodermic needle.

243

Sense	Number of Sensory Cells	Number of Afferent Neurons	Total Neuronal Channel Capacity (bits.s⁻¹)	Psycho-physical Channel Capacity (bits.s⁻¹)
Hearing	3.10^4	2.10^4	10^5	30
Smell	7.10^7	10^5	10^5	1
Taste	3.10^7	10^3	10^3	1
Touch	10^7	10^6	10^6	5
Vision	2.10^8	2.10^6	10^7	40

Alternatively, indirect external measurements are usually made by sphygmomanometer via a blood pressure cuff and stethoscope.

SPECIFIC COMMUNICATION - NERVOUS:

Information is measured in bits (binary digits), where one bit is equal to the information transmitted by a single binary symbol (e.g. yes/no, on/off, 0/1). When processing information the body as a whole has a sensory uptake of 10^9 bits.sec^{-1} and a motor output of 10^7 bits.sec^{-1}, but can only recognise 10^1-10^2 bits.sec^{-1}. The nervous system has a complex network of inhibitory and excitatory links between neurons by synapses, for encoding, decoding, and recoding signals. The predominant organ that co-ordinates processes within the body is the brain. This centre for communication is small but impressively complex. The adult brain weighs about 1.5 kg, equivalent to only 2.1 % of body weight (NB mammal brain weight = 0.122 x (body weight)$^{2/3}$). In the human brain there are about 25 x 10^9 neurons and 10^{15} synapses between them. This results in a phenomenal and impossible to comprehend number of possible combinations for the connections in the brain (cf. the universe, which is thought to consist of 10^{11} galaxies containing a total of 10^{22} stars).

It is possible to represent the various parts of the body according to the proportion of neural activity attributed to them. This is typically done by correlating the functional organization along the central gyrus of the cortex with the surface area of the body to which it corresponds. Such a representation is called a homunculus. The motor and somatosensory homunculi are usually considered separately, but reveal very similar results. The parts of the body are not represented in the cortex in proportion to their size. A disproportionately large area of cortex is concerned with controlling the face (especially the lips), hands (especially the fingers), and to a lesser degree the feet. Thus a caricature is created that provides a distorted map of the peripheral surface of the body. Those parts of the body with the most degrees of freedom for movement require more motor control for manipulation. Similarly, those parts of the body with the densest innervation produce more sensory input for spatial resolution. Motor and somatosensory homunculi resemble each other so closely because the parts of the body with the greatest dexterity also have the greatest sensitivity.

245

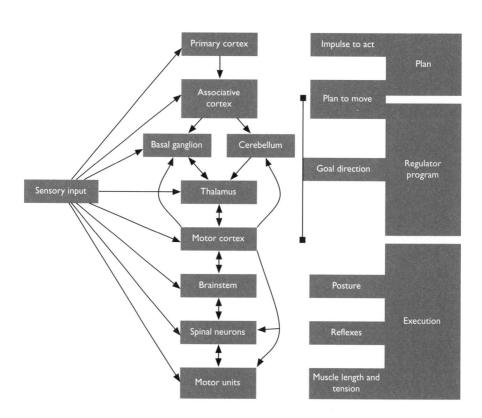

Neuronal connections (connexions):

Nerve impulses occur in specialized cells called neurons (nerve cells) that can undergo excitability. The simplest nervous arrangement is that of the reflex arc. In the reflex arc an impulse is transmitted from sensory receptor to motor effector via three neurons. These three neurons are connected in series, one after the other, via synapses. With the exception of these synapses all neurons are insulated from each other by a fatty sheath of myelin. Each neuron has an afferent fibre (dendron) bringing the impulse to the cell body, and also an efferent fibre (axon) retransmitting the impulse away from the cell body. The three neurons are in turn called a sensory neuron, relay neuron and motor neuron. This simple reflex arc is responsible for rapid involuntary responses to stimulation; such as blinking when an object approaches the eye, or constriction of the pupil in bright light. For slower voluntary control and integrated involuntary actions, more neurons are connected, both in series and parallel, to form a complex arrangement of synaptic interactions. One receptor can be connected to many effectors, or many receptors can be connected to one effector. In this way, one neuron may have up to 100,000 connections. Hence the arrangement of neurons into complex interconnected branching pathways can cause the divergence and convergence of signals. Neurons relay impulses to brainstem nuclei or brain centres; these are areas of neurons that share a common functionality. From these areas, impulses can be relayed even further to other nuclei or centres; for comparison with other impulses, or rather, the activity at other synapses found there. Different pathways can be excited or inhibited depending on the presence or absence of additional signals, and also the type of nerve fibre and distance between neurons and hence the timing of signals arriving at particular junctions. This all enables a highly complicated functional integration. Perhaps this is best exemplified by the ability to store and retrieve the information (memory) that is required for learning. The storing of information is linked to many processes, probably revolving around the perpetual cycling of patterns of excitation by networks of neurons in the hippocampus. One such process that memory is known to depend on, is the removal of proteins that accumulate around neurons (cf. formation of β-amyloid plaques in Alzheimer's disease), and this process is mediated by metalloproteinase-9 (MMP-3, the so-called memory enzyme).

Impulse transmission along neurons:

Action Potentials are point measurements of the conduction of impulses

247

Learning:

Interindividual communication is essential for the development of co-operation, and depends on learning. Learning is the process by which behaviour is modified on the basis of experience. Non-associative conditioning involves a change in response to a repeated stimulus (while associative conditioning requires learning a relationship between two events). A practical problem with measuring non-associative conditioning is that successive stimuli at arbitrary intervals will have a variable effect on reinforcement. In other words, differences in the duration between stimuli or the number of stimuli, will have an effect on the memory of those stimuli. It is essential to standardize measurements in order to directly compare different types of non-associative conditioning, and further to see if they can be combined to produce examples of more complex associative conditioning. It is therefore necessary to have a transferable measurement which is not influenced by the repetition of stimulus-response that seems to be demanded by the definition of conditioning. One solution would be to calculate the second measure of response as a percentage of the first measure of response, and plot this against the duration in time between those two measures. It is then only necessary to make one repetition and to deliberately vary the duration between the two stimuli (so as to cover a range of intervals) with successive independent stimulus-response events in different individuals. The main types of non-associative conditioning are habituation and sensitization. Habituation is where the stimulus has no consequence and so the response diminishes with time when the same stimulus is repeated and recognised. For example, startling does not sensitize otherwise energy would be wasted. If the stimulus becomes consequential then dishabituation occurs. Sensitization is where the stimulus has consequence and so the response increases with time when the same stimulus is repeated and recognised. For example pain, whether sharp to elicit a rapid reflex or a continuous ache to elicit a protective response. Pain does not habituate otherwise damage would be ignored. If stimulation decreases then desensitization occurs.

along neurons (NB conduction also occurs through myocardial cells). These occur by the rapid reversal of membrane polarity, whereby the cell exterior becomes temporarily more negative and the cell interior becomes less negative. The electrical potential across the membrane drops in negativity from the resting potential until it is slightly positive. This occurs due to a rapid influx of sodium ions. Following the cessation of sodium influx there is a complementary efflux of potassium ions to restore the original charge differential. Shortly thereafter the balance of ions is returned to the pre-excitation state by the continuously active sodium-potassium ATPase pump. Hence, propagation along neurons occurs following a successive 'domino' effect of sodium influx. Local currents at the leading edge of the change in charge facilitate self-propagation. The nerve impulse is not an electric current but a wave of change in potential. Thus the energy for an impulse does not come from the stimulus but from the nerve itself. Hence the velocity of the conduction is independent of the intensity of the stimulus.

Conduction velocity can only vary with different types of nerve fibre. Of successively slower velocities are: A fibres (large myelinated somatic), B fibres (small thinly-myelinated visceral) or C fibres (small and non-myelinated). Velocity is greater in myelinated fibres and is directly proportional to axon diameter. The largest diameter A fibres are 15 μm across, and have the fastest conduction at 160 m.s^{-1}. In addition, the conduction of an impulse either occurs or it does not, there is no variation in the degree of transmission. Whether transmission occurs or not in the first place can only be influenced by the condition of the nerve. For about 1×10^{-3} seconds following conduction there is an absolute refractory period during which no second impulse can be transmitted. Then for about 3×10^{-3} seconds following the absolute refractory period there is a relative refractory period during which the threshold for stimulation is elevated. This is time needed for physical recovery of the difference in electrical potential, but it also ensures that there is no overlap of signals. A large number of action potentials can be activated in rapid succession because each action potential only utilizes about 1/100,000th of the total ions available, and so is not immediately limited by the rate of active sodium-potassium pumping. An intense stimulus simply results in either an increase in the impulse firing rate, or activation of additional sensory neurons that have a higher threshold (recruitment), or both. The frequency of impulses required for a sustained contraction (tetanus) can vary depending on the muscle

FOR WRITTEN NOTES:

249

involved, from 10 impulses.s^{-1} for postural extension to 350 impulses.s^{-1} for eye movements.

Synaptic transmission across gap junctions:
For a signal to continue from one neuron to the next, the extracellular space between the two cells must be crossed. The end-plate regions of dendrites from neuron bodies are closely invested with the synaptic knobs at the end of axons from other neurons. Here the most efferent end of the impulse carrying neuron is separated from the most afferent end of the next neuron by an insulating gap junction (synaptic cleft). These opposing synaptic terminals are about 0.5 μm in diameter and separated by some 10-40 nm. Transmission across a synapse requires 20-30 vesicles (each about 30 nm in diameter) to each release 5,000-10,000 molecules of neurotransmitter (e.g. acetylecholine, epinephrine, dopamine) into the gap junction. The release and diffusion of the neurotransmitter results in a synaptic delay of 0.5-1 ms. The time taken to completely recycle each vesicle is 40 s. With repetitive stimulation this delay could lead to the fatigue of nerve transmission.

The Nervous System and Senses:
The nervous system can be sub-divided into several categories. There is the Autonomic Nervous System (ANS) which is comparatively slow, the Peripheral Nervous System (PNS) which is comparatively fast, the Central Nervous System (CNS) which has overall control, and the special sense organs which mediate sensation. Classically the special sense organs are the: retina for sight, cochlea for sound, various mechanoreceptors in the skin for touch, gustatory cells for taste and olfactory cells for smell. Importantly in addition to these, there are the: vestibular labyrinth for balance, nociceptors for pain, various proprioceptive sources (e.g. muscle contraction, skin stretch) for joint position, cold and warm thermoreceptors for temperature, and various visceroceptors for internal state (e.g. fullness of stomach and fullness of bladder). Many of these parameters are integrated, so that the different senses can function interdependently and reinforce each other. The approximate limitations and degrees of resolution for each special sense are as follows.

Vision: The range in contrast has a ratio of $1:10^3$ from lightest to darkest. The range in brightness (luminance or lightness) is $1:10^{11}$

For written notes:

251

from 10^{-6} cd.m^{-2} (overcast night sky) to 10^7 cd.m^{-2} (bright sunshine). Maximum resolution is at about 0.25 m (the near point) and is equivalent to 1/60th of a degree (i.e. between 10th-100th of 1 mm, or about 300 pixels.cm^{-1}). It is possible to discriminate 7 million colour valencies (different combinations of hue, saturation and lightness) within the frequency range of 400-750 nm. The angular velocity of the eye can match the movement of a fixated object up to 80^0.s^{-1}, and fine movements of the eye can be at up to 150 Hz (cf. the average computer screen has about 30 pixels.cm^{-1} and a refresh rate of 50 Hz).

Hearing: The ear can hear intensities from -20 to 140 dBSPL, and over frequencies from 20 Hz to 20 kHz. The best discrimination possible is about 1 dB in intensity, 3 Hz in frequency and 10^{-5} s in time (equivalent to localizing a sound to within 3^0 with binaural spatial orientation).

Touch: Best possible two-point discrimination is 3 mm (on the tongue), best threshold for single point stimulation is 5 mg (on the nose), and best threshold for skin deformation is 1.6 mm.s^{-1} (around the eye).

Taste: The qualities of flavour and their best thresholds for detection are salty at 0.01 mol.l^{-1} (sodium chloride), sour at 10^4 mol.l^{-1} (hydrochloric acid), sweet at 10^5 mol.l^{-1} (saccharin), and bitter at 10^6 mol.l^{-1} (quinine sulphate).

Smell: The classes of odour are: camphor (e.g. 1,8-cinole), ethereal (e.g. 1,2-dichloroethane), floral (e.g. d-1-β-phenylethylmethylcarbinol), musky (e.g. 3-methylcyclopentadecan-1-one), pungent (e.g. formic acid), and putrid (e.g. dimethylsulphide). Olfactory detection thresholds can be as low as 10^7 molecules.ml^{-1} in air (i.e. for butyl mercaptan from skunk musk).

The Autonomic Nervous System:
The ANS overlaps with, and has a vital controlling influence on, many other systems. The ANS is a branch of the nervous system controlling many involuntary functions. Thus it is *automatic* with no conscious control. The spinal nerves act independently of cerebral influence to control the internal environment. However the actions of the ANS on the whole body are co-ordinated by the medulla oblongata and hypothalamus. The ANS is particularly important for maintaining homeostasis and the 'fight or flight' response. For example acute stress

253

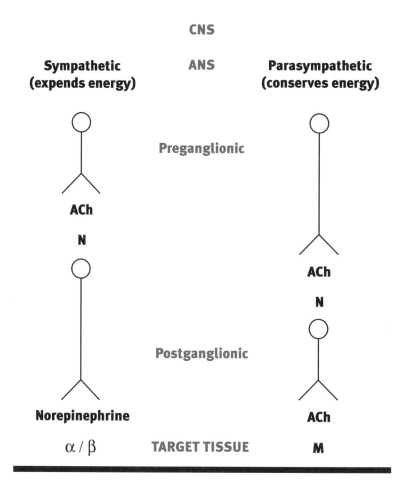

can cause an 'adrenaline rush', where the adrenal gland dramatically secretes more epinephrine (adrenaline!) which increases the rate and force of cardiac contractions and diverts blood from the skin and gastro-intestinal tract to the skeletal muscles, in preparation for exercise. The neurons of the ANS only carry information from the brain and so are described as *efferent*. One neuron in the outflow tract has its cell body outside the CNS in a *ganglion*. The *pre*ganglionic fibre runs from CNS to ganglion, whereas the *post*ganglionic fibre runs from ganglion to tissue. Preganglionic fibres are myelinated with a conduction velocity of 2-14 m.s^{-1}. Postganglionic fibres are unmyelinated with a conduction velocity of about 2 m.s^{-1}. There are both sympathetic and parasympathetic divisions. These two divisions complement each other, when one increases activity the other decreases activity. In general the sympathetic division expends energy, while the parasympathetic division conserves energy. In the sympathetic division preganglionic axons leave the CNS in the *thoraco-lumbar* region. Most ganglia are close to the CNS in the 'sympathetic chain', although a few are further away such as in the apron of tissue attached to the gastro-intestinal tract (mesentery) or in the neck. In the sympathetic division *pre*ganglionic fibres are *short* and *post*ganglionic fibres are *long*. The ganglia are remote from the target tissue, with most tissues being innervated to some extent. In the parasympathetic division preganglionic axons leave the CNS in the cranio-sacral region. In the parasympathetic division *pre*ganglionic fibres are *long* and *post*ganglionic fibres are *short*. The ganglia are either near or actually in the innervated tissue, with only selective tissues being innervated.

In the nervous system, signals are relayed across synaptic clefts, at the interface between different neurons. There is chemical neurotransmission at both the ganglion and neuroeffector junction, where neurotransmitters released from the presynaptic cell act at receptors on the postsynaptic cell. In the sympathetic division of the ANS, *pre*ganglionic fibres are cholinergic releasing *acetylcholine*, which acts at *nicotinic* receptors on postganglionic cells. Nicotinic receptors are classically stimulated (causing muscular contraction) by nicotine from the tobacco plant *Nicotiana tabacum*, and blocked (causing neuromuscular blockade and hence paralysis) by tubocurarine from the vine *Chondodendron tomentosum*. Whereas, *post*ganglionic fibres are adrenergic releasing (mainly) *nor-epinephrine*, which acts on α *and* β *receptors* on target cells. In the parasympathetic division of

FOR WRITTEN NOTES:

255

the ANS, both preganglionic and postganglionic fibres are cholinergic. *Pre*ganglionic fibres release *acetylcholine*, which acts at *nicotinic* receptors on postganglionic cells. Whereas, *post*ganglionic fibres release *acetylcholine*, which acts on *muscarinic* receptors on target cells. Muscarinic receptors are classically stimulated (causing bradycardia, increased secretions and fluid loss; and miosis, i.e. pupillary constriction) by muscarine from the 'magic' mushroom *Amanita muscaria*, and blocked (causing tachycardia, decreased secretions and fluid loss; and mydriasis, i.e. pupillary dilation) by atropine from the Deadly Nightshade plant *Atropa belladona*).

The ANS regulates secretions and fluid loss by altering: defecation (gastro-intestinal tract motility), micturition (urination or evacuation of the bladder), emesis (vomiting), salivation, sweating and lacrimation (production of tears). The ANS also contributes to: the control of blood pressure (by altering heart rate and vasomotor tone), the response to haemorrhage and shock, sexual excitation (involuntary arousal is under parasympathetic control, while orgasm is under sympathetic control), visual acuity (by altering pupil diameter), skin wrinkling on water immersion (facilitating an improved grip) and thermoregulation.

SPECIFIC TRANSPORT - MUSCULAR:

Muscle accounts for some 40-50 % of body weight. All muscles contract by converting chemical energy into mechanical energy and heat. In the muscular system there are three categories of muscle: skeletal muscle which in conjunction with the endoskeleton facilitates external movement, smooth muscle which facilitates internal movement (e.g. peristalsis; which is propulsion through a tube due to alternating contraction and relaxation, particularly in the gastro-intestinal tract), and cardiac muscle which facilitates a continuous circulation.

Microscopic mechanism of contraction:
Skeletal muscle has a number of functional properties that facilitate movement. Skeletal muscle is striated in appearance and under voluntary control. Each muscle cell (muscle fibre) contains a bundle of myofibrils, and each myofibril exhibits a banding pattern (striations) that is composed of a number of myofilaments. There are light thin bands of actin myofilaments, alternated with dark thick bands of myosin myofilaments. Alternate layers of actin myofilaments and

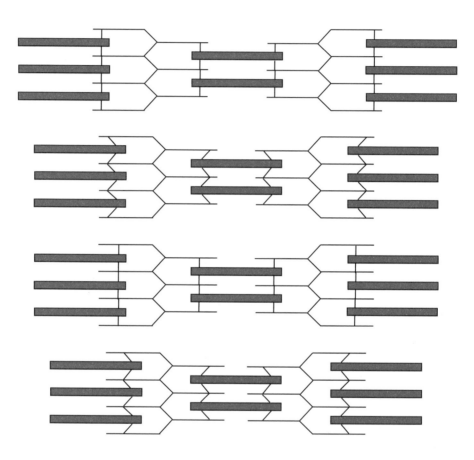

257

myosin myofilaments are orientated in parallel with the longitudinal axis of the muscle. The orientation is such that there are also alternate transverse bands. The longitudinal distance separating one band of actin from the next adjacent band of actin encompasses the functional unit of one sarcomere.

The shortening of skeletal muscle cells for contraction depends on the interaction of actin and myosin. During contraction the overlap between actin and myosin increases and the light band narrows, during relaxation the overlap decreases and the light band broadens. The myofilaments are effectively *sliding* over each other, coming closer together during contraction and further apart during relaxation. The length of the filaments is unchanged, only the degree of overlap changes. During sliding each sarcomere can contract to a minimum overall length of about 1.65 μm, or extend to a maximum overall length of about 3.65 μm. Actin-myosin bridges (cross-bridges or cross-links) connect the actin to the myosin and pull the myofilaments past one another by a ratchet mechanism. The ratchet mechanism involves a cycle of attachment, contraction, detachment and reattachment of the actin-myosin bridge. The contraction is brought about by a change in molecular configuration that results in a shift in angle of the actin-myosin bridge from 90° to 45°. This change in angle for each bridge requires one ATP molecule, which is supplied by mitochondria (sarcosomes) located between the myofilaments. Each cycle of change in bridge angle increases the overlap of actin and myosin by 8-10 nm at each end (a cycle must include both ends for contraction to occur). This is equivalent to a shortening of each sarcomere by about 1 %. Thus one cycle in every sarcomere will contract the whole muscle by about 1 %. During contraction these cycles can occur 50-100 times each second in every sarcomere. Therefore a 50 % reduction in muscle length can be achieved in 0.5-1 s.

Excitation-contraction coupling within muscle depends on the action of calcium and actin-myosin bridges. The arrival of an action potential leads to the release of acetylcholine that activates the receptor at the postsynaptic neuromuscular junction. This leads to the depolarization of the muscle cells and influx of calcium ions. Calcium ions are stored in the endoplasmic (sarcoplasmic) reticulum that envelops each of the myofibrils within each muscle cell. The cell membrane (sarcolemma) invaginates between the myofibrils in a T-system of transverse tubules.

For written notes:

259

A triad is a location where two opposing reticuli are separated by an invagination of the cell membrane (i.e. where there is a sarcoplasmic reticulum on either side of a sarcolemma T-tubule). This combination of reticuli and membrane enable an even distribution of calcium ions throughout the cell. Within each myofibril the calcium enables troponin to change the configuration of tropomyosin with respect to F-actin. Tropomyosin blocks the formation of actin-myosin bridges, so removal of tropomyosin enables the actin-myosin bridges to form and ATP to be used for contraction. At the end of the contraction ATP also breaks the actin-myosin bridge (cf. lack of ATP causes *rigor mortis*; which stiffens skeletal muscle within 4 hours of death, and lasts until broken by cellular autolysis 24-48 hours after death). The cycle of actin-myosin bridging can then be repeated in a rapid ratchet-like movement, up to a limit dependent on the extent of calcium influx. The increase in overlap of actin and myosin occurs for many myofilaments in parallel *and series*, in order to bring about the contraction of a muscle.

Macroscopic musculoskeletal transportation (biomechanics):
Skeletal muscle acts on bone to enable locomotion. In the human adult there are over 600 muscles, attached to 206 bones (54 of these bones are in the hands and 44 in the feet). Locomotion is the movement of the whole body through the environment, this is controlled by behavioural choices in order to optimize conditions. The musculoskeletal system relies on levers for the movement of the body by locomotion. For musculoskeletal mechanics to work, muscles are anchored to bones via tendons. During contraction muscle cells get shorter and fatter to pull body parts together. Muscles can't expand only relax, so muscles 'pair' in opposition either side of joints. In most cases joint friction is minimised by synovial fluid which acts as a lubricant. Effort of contraction is used to move either internal or external loads. This effort can either be closer to the joint (ie. fulcrum) than the load, and the distance moved by the effort is smaller than that moved by the load, or the effort can be further from the fulcrum than the load, and the distance moved by the effort is then larger than that moved by the load.

Skeletal bone has a number of functional properties that facilitate movement. Bones are comparatively rigid and can be articulated where they join. Thus joints provide the opportunity to move whole parts of the body with respect to the rest of the body. In this way, the bones in the body act as levers that are operated by the muscles. There are

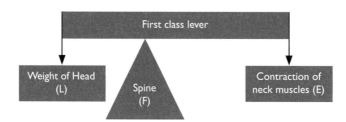

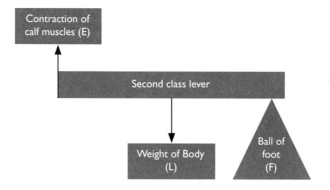

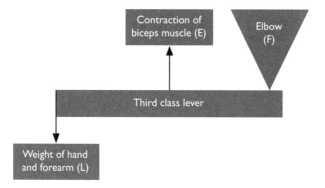

three main types of lever (first, second and third order) depending on the positions of the pulling force, the weight which is being lifted, and on where the lever is pivoted. The weight lifted divided by the force is called the mechanical advantage (MA):

MA = Load (L) / Effort (E).

Most muscles work at a mechanical disadvantage (MA <1) because the pulling force is exerted closer to the fulcrum than where the load is. For a *First class* lever MA can be <1 or >1. An example of such a lever is the support of the head by the neck, where the pivot is between the weight and the force (L = weight of head, E = contraction of neck muscles, and F = atlas vertebra). For a *Second class* lever MA >1 and E < L. An example of such a lever is active plantar flexion, where the weight is between the pivot and the force (L = weight of body, E = contraction of calf muscles, and F = ball of foot). The longer the heel, the greater the mechanical advantage with which the calf muscles work. For a *Third class* lever MA <1 and E > L. An example of such a lever is active elbow flexion, where the pulling force is between the pivot and the weight (L = weight of forearm, E = contraction of arm muscles, and F = elbow joint). Most levers in the body are third class. The different levers of the many joints in the body cover a wide range of possible forces and actions. Actions are either static (i.e. support against gravity) or dynamic (i.e. locomotion). Static actions are reactive (e.g. to push away, maintain posture and balance), and the forces involved include: compression, tilt and shear (e.g. at the end of long bones or vertebral discs). Dynamic forces include: bending and lifting (e.g. to pull towards or to pick up); variable friction (e.g. slipping of a foot on the ground); rotation (e.g. generating torque at the hip and knee when walking); and dynamic cycles (e.g. of the legs when running).

Dynamic forces encompass a wide range of opposing movements. Examples include: flexion (decreasing joint angle) and extension (increasing joint angle), internal rotation (inward roll. a specialized form of which is pronation), external rotation (outward roll, a specialized form of which is supination), abduction (away from body), adduction (towards body) and opposition (diametrical movement of the thumb towards any finger of the same hand). These methods of pivoting forces around a fulcrum are essential for locomotion. Thus the basic control of muscle and bone allows specific transportation of parts of the

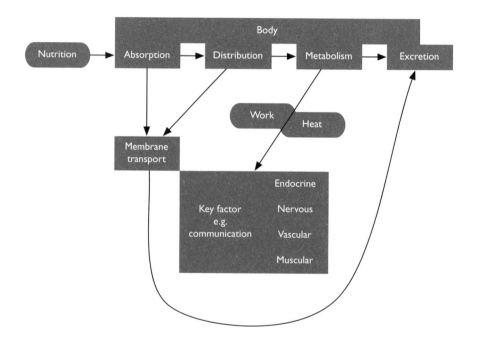

body with respect to other parts, or many joints can work together to transport the whole body. This provides the ability to physically interact with the environment by moving through the environment. Maximizing the efficiency of the body depends on a balance between supply and demand. This optimization of resources requires the fulfilment of drives for attraction and avoidance, and these drives depend on a means of transporting the body. Therefore, locomotion is important to maintain the life of the body as a whole.

SUMMARY OF COMMUNICATION:

Blood and muscle may provide means of transport, but it is nerves and hormones that are *the* means of communication in the body. Both nerves and hormones employ a wide range of signals, but all signals have common characteristics. In order to regulate many diverse processes it is important that signals are insulated from each other. However, for regulation of individual processes it is equally important that related signals can interact. Indeed at some level all processes must be related, and so the degree of signal interaction will vary according to the disparity of the processes in question. When any one signal interacts with another signal, it can have one of four possible effects. These effects are additivity, antagonism, potentiation and synergism. Addition is simply a summation of the effect that would have occurred if the two signals were separate. An antagonistic effect results in less effect than if the two signals were separate; there is an opposing and hence inhibitory relationship (cf. an agonistic effect activates changes). Potentiation occurs when one signal that would not by itself have an effect, augments the effect of another signal. A synergistic effect results in more of an effect than simple summation of two additive signals would allow. In turn, these four possibilities for direct interaction can each occur due to: a functional feedback mechanism, a chemical change in activation, a difference in disposition (referring to absorption, metabolism, distribution and excretion; ADME) or a specific receptor relationship.

FOR WRITTEN NOTES:

265

OVERALL SUMMARY:

In essence the focus here is on the things that are done, rather than on what does the doing. There is an underlying emphasis on the effects of deficit and excess. There are many things in a living body that must exactly match, such as energy production and consumption, or water gains and losses. What is of key importance is maintaining the balance of these many interdependent parameters at the same time. Trying to understand the integration of the processes involved poses more questions than answers. For example, it is relatively easy to explain how increasing urine output decreases body volume, hydration, acidity, temperature and energy. But how do all these things - and many more - interrelate to perpetuate life. Every time it appears that a comprehensive view is just about to come into focus, so it is blurred by the realization of how much more there is to understand, and thus how little is really comprehended. In other words, increasing understanding intrinsically increases how much there is to understand. Each little chip of hard work produces more questions than answers. Rarely it is possible to pull these divergent strands of work together, converging ideas into a leap of understanding. This is the legacy of successful science. Remember that successful science depends on having ideas and testing them thoroughly. This requires inspiration and dedication. In this respect it is the measurements that we consider important that shape our perception of the world. What amazing creatures we are, that we can detach ourselves from the process of living in order to contemplate the living process.

266

COMMON MEDICAL CONDITIONS

Arrest
      Cardiac (cf. ventricular arrhythmias)
      Respiratory (cf. tension pneumothorax)
Arthritis (osteo and rheumatoid)
Asthma (and other allergies)
Cancer (malignancy)
Congenital malformation (acquired and genetic)
Chronic Obstructive Pulmonary (Airways) Disease
Diabetes (insipidus and mellitus)
Degenerative disease (e.g. Alzheimer's, Parkinson's)
Epilepsy
Gastro-intestinal obstruction
Heart disease (e.g. angina)
Infectious disease
Malnutrition (e.g. obesity)
Mental illness (e.g. depression)
Oedema
Poisoning (over-dosage)
Trauma (e.g. fracture, burns, haemorrhage)
Shock (e.g. cardiogenic, hypovolaemic, or peripheral blood
      sequestration; the latter including bacterial,
      pharmacological and spinal cord injury as causes)
Vascular disease (e.g. coronary thrombosis, dissecting aortic
      aneurysm, pulmonary embolism, stroke, varices)

ADDENDUM ON MODERN MEDICINE

A HOLISTIC APPROACH TO MEDICINE

There are a wide variety of different complementary medicines from homeopathy to acupuncture. In general the common underlying theme is that of a whole body (holistic) approach. This is underpinned by attempts to restore the body to a state of *balance* and maintain it there (e.g. Ying and Yang). This is in contrast to the 'system' specialization of conventional modern medicine. Of course with the depth and breadth of current knowledge, specialization is a practical necessity in order to master the skills required and maximize clinical effectiveness (e.g. to perform a kidney transplant with the best chance of success). However, this must be done without neglecting potential repercussions for the rest of the body. The great asset of complementary medicine is the whole body ethic. Unfortunately this is often gravely undermined by a lack of acceptance of other therapies. Each type of complementary medicine claims in its own way to provide a complete approach to treatment. It is this exclusivity and jealous picketing of individual boundaries that harbours danger for the patient undertaking complementary therapy.

Complementary treatments that work have always been incorporated into conventional treatments. Complementary therapies always have inadequate evidence of their efficiency. If a treatment was ineffective or caused more harm than benefit then it would be stopped. If a treatment was effective then it would be integrated into modern medicine. This is often unrecognized. The origins of conventional medicine are what was once part of what is now called complementary medicine. Thus, most of what would be useful has been subsumed already. Sometimes complementary treatments are acclaimed as a more natural therapy, when in actual fact an identical substance is "peddled" under a generic name. Complementary treatments for which there is no evidence of effectiveness often exploit differences in psychological approach and economics. In other words, the paying 'customer' receives more counselling time during which personal beliefs are reaffirmed. This increases the probability of a mind over body (psychosomatic) response. In these cases there is no physical link to the treatment, i.e. it has a placebo effect. This is accentuated by a higher use of complementary medicine by individuals for which conventional medicine has 'failed' or can provide little hope. Pain drives people to seek relief. Pain has both physical and emotional components, and thus has a

COMMON CLINICAL MEASURES (AND THEIR ABBREVIATIONS)

Angiography
Biological half-life ($t^1/2$)
Blood pressure (BP)
Creatinine clearance (CC)
Central venous pressure (CVP)
Computerized tomography (CT)
Doppler
Echocardiogram (Echo)
Electrocardiogram (ECG)
Electroencephalogram (EEG)
Electromyogram (EMG)
Erythrocyte sedimentation rate (ESR)
Full blood count (FBC)
Forced expiratory volume in one second (FEV1)
Forced vital capacity (FVC)
Glasgow Coma Scale (GCS, a score of consciousness level)
Glomerular filtration rate (GFR)
Intracranial pressure (ICP)
International normalized ratio (INR, or prothrombin ratio)
Jugular venous pressure (JVP)
Kaolin cephalin clotting time (KCCT)
Liver function test (LFT)
Mean cell volume (MCV)
Nuclear magnetic resonance (NMR)
Packed cell volume (PCV, or haematocrit)
Positron emission tomography (PET)
Prothrombin time (PTT)
Radiograph (not X-ray)
Red blood cell count (RBC, RCC or erythrocyte count)
Temperature (^{o}T)
Temperature, pulse and respiration count (TPR)
Ultrasound scan (USS)
Urea and electrolytes (U&Es)
Ventilation-perfusion ratio (V/Q)
White blood cell count (WBC or WCC)

subjective aspect which involves complex psychological interactions. Pain can be a manifestation of other social problems, and there can be much imagined that is conditioned. There can also be social pressure to conform to the success of alternative complementary 'cures' (as the sick role-model achieves attention, money spent is vindicated, and socially undesirable mental anguish is transformed into a more socially acceptable physical form). Desperate people need something to believe in. Unfortunately, a few treatments are even directly dangerous. However most are only indirectly dangerous, by application instead of - or interfering with – conventional treatments that are known to be effective. Thus a measured certainty of success, is replaced with a certainty that is not only unmeasured, but is potentially harmful.

It is true that the mechanisms of action for the vast majority of conventional treatments are not fully understood. However, all conventional medical treatments have evidence of effectiveness at least in the form of unbiased reproducible success. All complementary approaches must submit to the rigors of scientific testing (i.e. randomized double-blind controlled trials) and be either integrated with modern conventional medicine or discarded. The best of everything needs to be extracted and combined in order to provide the most effective treatment. The superstition of today may well be the biomedical science of tomorrow. But we must have evidence to justify diverting resources, belief is not enough.

REFERENCES

An evidence base should be provided to support various points made in the preceding text. In particular I would like to include primary references for all cases of quantification. However this would unnecessarily overburden readability and is omitted to facilitate a greater ease of understanding. I unreservedly apologize for not giving credit to those to whom it is due.

Furthermore, I would not advocate a specific bibliography; as books that I find useful may not suit you. I would recommend you select one hefty tome to act as a comprehensive reference. In all probability you will never read such a tome from cover to cover but you will frequently use it for 'looking things up'. So before buying, test that the index is extensive and has the obscure entries you are likely to require. Additional books should be selected on the basis that you will read at least complete chapters from them. Choose books that you find have both an enjoyable format and a flowing writing style. In short, pick books you think you can 'get on with' and are likely to actually read. This will be more use to you than an abandoned 'recommended text'. Do not pick books that you will never read and have only selected in the false hope of 'learning by diffusion' because of ownership and close proximity. Finally, whatever you do read, remember the fact that all books contain mistakes.

A PHILOSOPHICAL CRUX

Don't believe everything you read in books, don't believe everything you're told by teachers, and don't believe everything you do is right. Be critical for nothing is certain. Logically challenge received wisdom, as it is the advancement of understanding that matters, and there is always an alternative point of view. The fact that someone else is right doesn't mean that you are wrong. Perhaps both points of view are right, perhaps both points of view are wrong, or perhaps both are right and wrong. Having this approach and always looking for alternative explanations that oppose your own, increases the depth and breadth of understanding. That may well be, but be that as it may, our duty is clear, truth.

INDEX

274

G

J

K

L

N

T

292

X

Y

Z

293

Time is the best teacher;
unfortunately it kills all its students.
adapted from Hector Berloiz (1803-69).